IEE CONTROL ENGINEERING SERIES 36

Series Editors: Prof. D. P. Atherton
Dr K. Warwick

ROBOT CONTROL

Theory and applications

Other volumes in this series

ROBOT CONTROL

Theory and applications

Edited by

K. Warwick and A. Pugh

Peter Peregrinus Ltd on behalf of the Institution of Electrical Engineers

Published by: Peter Peregrinus Ltd., London, United Kingdom

© 1988 Peter Peregrinus Ltd.

ISBN 0 86341 128 2

Printed in England by Short Run Press Ltd., Exeter

Contents

List of authors

1   P. Adl and R.T. Rakowski; Brunel University, U.K.

2   P.C. Wong, M.T. Thorn and S, Littlewood; Hunddersfield Polytechnic, U.K.

3   S.A. Carr and M.J. Grimble; University of Strathclyde, U.K.
     G. Anderson and J.V. Ringwood; National Institute for Higher Education, Dublin.

4   M. Farsi and K.J. Zachariah; University of Newcastle upon Tyne, U.K.
     K. Warwick; University of Warwick, U.K.

5   J.Z. Sasiadek and R. Srinivasan; Carleton University, Canada.

6   Z. Taha; University of Malaya, Malaysia.

7   A. De Luca, P. Lucibello and F. Nicolò; Università di Roma, Italy.

8   A.Y. Zomaya and A.S. Morris; University of Sheffield, U.K.

9   M. Dickinson; Sheffield City Polytechnic, U.K.
     A.S. Morris; University of Sheffield, U.K.

10   S.R. Pandian, M. Hanmandlu and M. Gopal; Indian Institute of Technology, New Delhi, India.

11   S.N. Ahmad and K. Natarajan; Nuclear Power Corporation, Bombay, India.

12   K. Ninomiya, I. Nakatani, J. Kawaguchi and K. Harima; Institute of Space and Astronautical Science, Tokyo, Japan.
     K. Tsuchiya, M. Inoue and K. Yamada; Mitsubishi Electric Corporation, Hyogo, Japan

14 J. Pu and R.H. Weston; Loughborough University of Technology, U.K.

15 L.G. van Willigenburg; Delft University of Technology, The Netherlands.

16 P. Chiacchio and B. Siciliano; Università di Napoli, Italy

17 S. Engell and A. Kleiner; Fraunhofer Institut IITB Karlsruhe, Federal Republic and Germany.

18 W.M. Grimm, N. Becker and P.M. Frank; University of Duisberg, Federal Republic of Germany

19 P.C. Müller and J. Ackermann; University of Wuppertal, Federal Republic of Germany.

20 R.A. Basta, R. Mehrotra and M.R. Varanasi; University of South Florida, Tampa, U.S.A.

21 J.E. Sanguino, J.S. Mata, A.P. Abreu and J.J. Sentieiro; Technical University of Lisbon, Portugal.

22 R.M. Crowder; University of Southampton, U.K.

23 A. Ghanem; San Francisco State University, U.S.A.

24 J.E.E. Sharpe; Queen Mary College, London, U.K.

Preface

This book offers a cross-sectional view of recent research and development carried out in the field of robot control. In, what one may regard as, a fairly well defined topic area, a large number of diverse requirements are thrown together in order to achieve a particular aim. The methods by which that aim is realised depend on such as the type of robot, its workplace, the workpiece spectrum, sensing elements employed, programmed control algorithms, the hardware base and software selected. The papers in this volume are wide ranging in their coverage of the general robot control area and present an up-to-date picture of the subject.

The book contains edited papers, given at the IEE International Workshop on 'Robot Control: Theory and Applications', held at the University of Oxford in April 1988. The main objects of the meeting being to enable the presentation and discussion of recent research results and novel developments. The Workshop was therefore allowed to have a broad scope, and the selection procedure applied to the papers submitted was carried out with this in mind. When given, the papers were divided into themed sessions as can be seen from the Contents section, which allows for each session to be regarded separately if desired.

The integration of sensory information into an overall control scheme is a increasingly important aspect in the development of a systems approach to robot control, and this must enable reactions, possibly in real-time, to a varying environment or requirements. Both tactile and vision sensing elements are considered in this text. Further, the controllers employed at each level must be able to quickly adapt to these changes such that appropriate commands can be efficiently issued. Various adaptive schemes are discussed, including self-tuning, model reference and variable structure techniques.

Response modelling is of primary importance in formulating and assessing suitable control strategies, such as those necessary for the close control requirements of manipulator trajectory tracking. Several aspects of modelling are considered and the means by which tracking can be carried out are shown in terms of links formed with parameter estimation schemes, as those employed within adaptive controllers.

A selection of manipulator implementation results are covered, ranging from direct usage in nuclear power plants, through space manipulators and walking robots, to high speed applications which necessitate fast sampling rates. The important topic of Robust Control is also discussed. Robust Control methodologies have been of great prominence in the control systems research arena over the last few years, and some of the results, which have been obtained, are shown, via several papers, to contribute particularly to robot control. The combined papers on this subject area provide an excellent overview of robust controller design, within the context of robot manipulators.

The overall efficient control of manipulator performance can only be achieved when its working environment is also taken into account. This means that if more than one manipulator is being employed, the multi-manipulator operation should be coordinated. Environmental aspects, however, can also include human operatives in several guises, and stipulated performance objectives should take these points into account, as discussed in the final papers.

In conclusion, the editors would like to thank all of those who have helped both in the organisation of the Workshop and in the preparation of this book. Particular gratitude is extended to Andrew Wilson at the IEE for smoothly coordinating planning and operation of the Workshop, and to John Billingsley and Richard Weston for acting as session chairmen. At Peter Peregrinus Ltd, our thanks go to John St.Aubyn and Nick Bliss for enabling a quick turnaround from receipt of final copy to publication. Finally, a special thank you goes to Marie Jones who dealt excellently with the necessary modifications.

Kevin Warwick
Alan Pugh

January, 1988.

Chapter 1

Automatic compensation system for adaptive gripping using a magnetoresistive force and slip sensor

P. Adl and R. T. Rakowski

1. INTRODUCTION

Intelligent robots will need sensors that will give them artificial senses of sight and touch, and so be capable of adapting to their environment. Achieving such a goal will open up new application areas for robot technology not only in the traditional engineering sector but also in areas such as the food processing industry (1). Using cameras and pattern recognition programs in robotic applications is well advanced. However, with the problem concerning the sense of touch, the few systems developed have concentrated on concepts leading to devices which are not environmentally robust nor allow sufficient flexibility of application. Since most tasks in the area of gripping and placing can be achieved by responding to contact and force, it is desirable to instrument the gripper with force and touch sensors (2).

In practice the gripper is usually constructed as an open-loop control system and therefore the object can be badly damaged due to excessively strong gripping forces. Such an inconvenience cannot always be avoided by feedback control of the grasping force because it is not easy to specify its optimum value. In view of the recent trend towards the extensive use of industrial robots for handling objects of unknown (or varying) characteristics, it is very important to determine the optimum grasping force by some adaptive means so that objects are not damaged or dropped. One of the most effective means of controlling the grasping force is to use the feedback signal from slip detectors.

2. SENSOR REQUIREMENTS

2.1 Types and Tasks of Taction

In specifying the type of task to be performed two functions for the sensor can be described: to detect object shape and to measure the forces determining the motion of the object relative to the hand. Because these two functions are completely different it is possible to speak of two different types of tactile sensor: a 'shape' and a 'manipulative' one. It is easy to argue that most industrial manipulation is not concerned so much with the use of tactile sensing for object identification since it is almost invariably done better by optical means. Thus it seems that the primary role for a tactile sensor is a manipulative one, i.e., one that monitors the dynamics within the hand.

In the simplest possible case a robot has a manipulator with two fingers, having internal surfaces planar and parallel to each other. Objects grasped between the fingers can make point, line or plane contacts with these surfaces. The need for sensing arises when trying to carry out manipulative tasks. During manipulation within the fingers the object is temporarily out of equilibrium. Monitoring of the forces applied by the fingers at the points of contact are required. Assuming that the object remains always in contact with the fingers, it is possible to restrict attention to the position of the object confined between two fixed planes. The object then has three degrees of freedom; it can slide or slip parallel to the finger planes and can rotate about the axes perpendicular to them.

2.2   Tactile Technologies

Considerable effort has been directed at developing a workable tactile sensor - both in industry and in the academic community (3). Diverse approaches have been explored to develop tactile transducers with sensitivity and robustness using various physical effects. Most workers have different ideas on what kind of information can best be gained through taction and what to do with it once it has been gathered. However, shifting the emphasis from shape recognition to force monitoring eliminates many of the physical effects which appeared promising for shape tactile sensing. If the three dimensional force vector and • associated couples are to be detected, then shear force sensing is vital. The limitation of measuring normal forces or pressures only rules out many tranduction technologies which have been proposed.

This limitation is primarily due to the nature of the transduction effect. In fact transduction is generally at the atomic level, where the transduction effect (e.g., piezo-resistive, strain gauge, piezo-electric, magneto-elastic, etc.) couples one form of energy to another. An inherent limitation of these 'microscopic' effects is the trade-off between sensitivity and fragility, and the restriction imposed by the design of the supporting structure on the direction of the forces that can be detected.

2.3   Using Magnetic Sensors for Force Detection

Of all the transduction techniques examined the only devices that have the capability of measuring the three dimensional force vector at numerous points in an array and be compact enough to be mounted on a robot gripper are certain magnetic field ones. Research work at Brunel (4) indicated that only Hall and Magnetoresistive effect devices have the ability to be fabricated in arrays, in small enough dimensions and compact enough to be implemented on a robot gripper. The magnetoresistive sensor, however, is more sensitive than Hall elements and can usually operate over a much wider temperature range. Moreover, its frequency range is much greater, since the magnetoresistive effect is not an inductive effect and can detect both d.c. and a.c. fields up to several megahertz. It is also a two terminal device and is not subjected to the many limitations inherent in a four-terminal Hall effect sensor.

3. MAGNETORESISTOR CHARACTERISTICS

3.1 Review

The magnetoresistive sensor is one of the more recent developments for detecting magnetic field variations (5). Magnetoresistive devices make use of a well known property of certain ferromagnetic materials such as permalloy (81% Ni and 19% Fe) to change resistivity in the presence of an external magnetic field. A thin, plannar single domain magnetoresistive film can be employed to sense a magnetic field to which it is exposed by passing an electrical sense current (either a.c. or d.c.) through the film, the films magnetisation vector being at an angle to the direction of current flow. The field being sensed exerts a torque on the magnetic moment in the film, causing the resistance of the film to increase or decrease depending on the sense and magnitude of the field applied to the film. The resistance of the film is therefore the analogue of the field strength. The exact theory of operation is well described by McGuire and Potter (6).

3.2 Fabrication

The cost of designing and manufacturing of magnetoresistive sensors for prototype devices is quite high. The method used for generating thin film patterns was to use evaporation masks. A suitably shaped aperture (referred to as a mask) has the desired pattern cut or etched into it. In a high vacuum chamber the mask is placed in close proximity to the glass substrate, thereby allowing condensation of the evaporant permalloy vapour only in the exposed substrate areas. The film thickness can be controlled by a rate monitor and a shutter.

Magnetoresistive sensors are the product of several evaporations. First a layer of Ni-Fe is deposited onto the substrate to form the permalloy element. This has to be done in the presence of a magnetic field aligned parallel to the plane of the film and in a direction related to the final magnetoresistive sensor geometry in order to provide a known magnetisation axis. Then, via a different set of masks, a layer of copper or gold is deposited at both ends of the element ot allow connection of the sensor to external circuits. Next an insulation layer, usually of photoresist is formed over the element and conductors to protect the

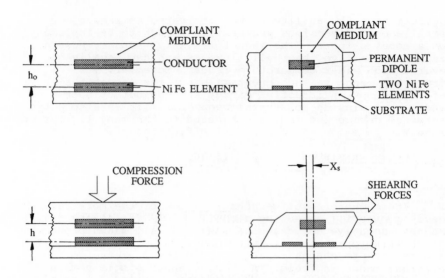

Fig.1.   Normal and shear force detection

sensor during subsequent use.

The sensor characteristics can be finely controlled by altering the permalloy element geometry and configuration. The sensor configuration details which were designed specifically for this project cannot be revealed due to a patent protection application.

3.3   Force and Slip Detection

No matter how adaptable its control system, how large its memory, or how many articulations it possesses, the capability of the robot is ultimately determined by how well it handles objects. It must be able to grasp, lift, and manipulate workpieces without causing any damage and without letting go. Clearly the grasping surface of a robotic gripper will also be the sensing surface for the magnetoresistive force sensor. The forces exerted by the gripper on an object will deform a compliant medium and will thus displace a current carrying conductor or a magnetic dipole (Fig.1) with respect to the magnetoresistor. The magnetoresistor will detect a change in magnetic field and produce an electrical signal. This signal will be proportional to the distance the elastic medium has been deformed and hence represents the shear or compressive displacement forces.

The choice and design of the gripping surface is very critical. Tactile transducers must be wear-resistant; they should be able to withstand abuse, particularly in rugged industrial environments. Of course, specialised applications can require resistance to extreme conditions. However, most conventional positioning, assembly and inspection operations require no unusual durability or environmental resistance

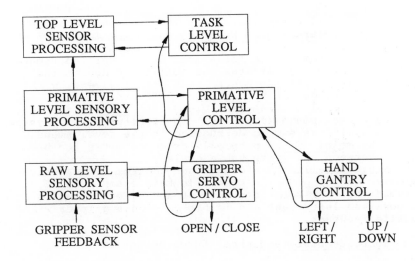

Fig.2.   Hierarchical control for a basic adaptive
         gripper application.

(7). There are probably a large number of applications in
which human hands may be directly replaced by automatic
hands. Bearing in mind these considerations, natural rubber
was chosen for the elastic medium offering a good balance of
properties, in particular for its excellent mechanical
characteristics.

The main conclusion that can be drawn from results to
date is the excellent repeatability of the measured values.
Undoubtably repeatability is a far more important factor
than any other, even accuracy. For although accuracy is
highly desirable, the primary requirement for indicating
trends must be to produce a consistent result continuously.

Finally, what is the best distribution of these
elements on a tactile sensor? To date grippers have always
had to be customised and consequently tactile sensors will
also need to be. Obviously, standardised devices and data
processing packages are indispensable. General purpose
design should reside mostly at the system level, while the
touch sensors are specially configured for the application.
To this end a modular design with linked sensing 'units' of
either normal or shear force transducers could be used to
construct the array. The optimum distribution of these
units might then be resolved for different tasks by careful
analysis of the application.

4.   HIERARCHICAL CONTROL

4.1 Advantages

The overall control strategy is based on the National Bureau of Standards (USA) Hierarchical Robot Control System (8) which allows for real time sensor interaction. This particular system allows for a modular standardised control technique, which facilitates the ability to add to the system or integrate the system with other systems, with the minimum of changes in hardware and software.
The control problem is divided into clearly defined modules based upon a simple IF/THEN structure. The system, therefore, allows simple de-bugging in the development states by means of the naturally produced sites for break points between the state table cycles.
Fig.2 details the hierarchical control for a basic adaptive gripping application. In this example it is assumed that the robot consists of a hand gantry which moves up/down and left/right and is fitted with a gripper with tactile feedback.

4.2 Software Implementation

Each control level cycles around the following procedures: INPUT PRE-PROCESSING; STATE TABLE IMPLEMENTATION; OUTPUT POST-PROCESSING. The input and output pre/post processing is mainly a data preparation task, the bulk of the processing being carried out by the state-table implementation as shown in Fig.3., where, depending upon the input conditions, one of the output conditions is executed.

4.3 Hardware Implementation

For adaptive gripping applications the use of a single processor is impractical due to the data processing requirements and response times. As a result a multiprocessor approach was developed (9). The system adopted used a separate single board computer for each of the control decision level blocks (i.e., a total of four boards).
The choice of an inter-processor communications link requires careful design. In order to facilitate system expansion in future applications the technique adopted used a common entered memory which would hold all the parameters which needed to be transferred between levels. Two viable solutions evolved which could manage the problems associated with a common memory bus. The problem of boards attempting to communicate at the same time could be solved by using a control clock to synchronise each state within the processors cycle around the pre-process/state-table/ post-process loop. However, this method produces a lot of wasted processor time. The technique adopted, therefore, was based on the principle of 'first come first serve'. Here as soon as one processor (CPU) attempts to talk to the

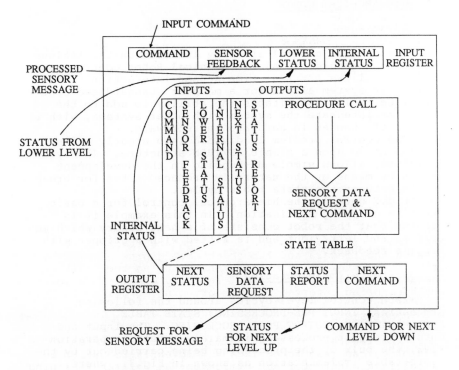

Fig.3.   Modified state table implementation.

common memory the other processors are immediately blocked
out. To avoid the problem of another board simultaneously
attempting to use the bus (resulting in possible loss of
data) the common memory unit sends an interrupt signal to
all the boards except the first to attempt to use the
memory. This technique avoids an overall cycle control and
the associated processor time wastage.

5.   ADAPTIVE GRIPPING

Since the overall control system is designed as modules
there is no reason why such a system could not be integrated
into an existing manufacturing system. When a pick command
is received control is decomposed down to the primary
command level and hence allows the slip and force sensor to
control the gripping process. The grip force control signal
for a simple application is shown in Fig.4. The voltage
signal provided to the servo-motor controller is directly
proportional to the required grip force. The first part of
the diagram shows a maximum -ve voltage (maximum motor
torque) during the 'open gripper' operation. The second
part of the diagram shows the gripper operating in 'adaptive
gripping' mode. Firstly, the gripper closes with a maximum
velocity. As soon as contact is detected by the tactile

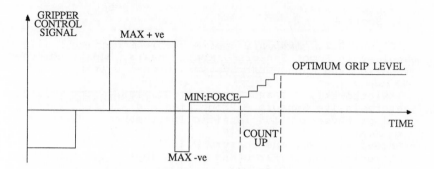

Fig.4.   Output from gripper drive circuit

sensor pad a short -ve pulse stops the gripper motor dead.
A minimum grip level is applied to ensure friction contact
between the fingers and the object being picked up. The
grip force level is then incremented during the 'lift'
operation until no slip signal is registered. This point
would then indicate the optimum gripping force.

6.   CONCLUSIONS

Experimental work has indicated that magnetoresistive
technology can lead to devices sufficiently robust to be
useful in a manufacturing environment as well as retaining
the flexibility to increase its resolution and increase the
range of application. Sensors could be tailor-made to a
particular application: grasp verification, contour
examination, gap detection and, of course, force and slip
detection. The use of hierarchical control strategy for
adaptive gripping has shown much promise and current
research is exploring the practical performance limits of
such a system. It is also possible to modify the gripper
control system to realise an adaptive assembly gripper based
on shear and normal force signals recorded during a
peg-in-the-hole assembly operation.

ACKNOWLEDGEMENTS

The authors are indebted to the U.K. Science and
Engineering Research Council for the support of the work.

REFERENCES

1.   'Food Manufacture using Robots and Flexible
Automation', Seminar, The Institution of Mechanical
Engineers, November 1987.

2.   Bicchi, A., Bergamasco, M., Dario, P., 1988,
'Integrated Tactile Sensing for Gripper Finger', Proceedings
of RoViSec-7, Zurich, Switzerland.

3.   Yardley, A.M., and Baker, K.D., 1986, 'Tactile Sensors for Robots: A Review', World Year Book of Robotics Research, 47-83, Scott P.

4.   Flynn, D.I., 1985, 'Use of Magnetic Recording Technology in Robot Sensors', MPhil Thesis, Department of Engineering and Management Systems, Brunel University.

5.   Kwiatkowski, W., and Tumanski, S., 1986, 'The Permalloy Magnetoresistive Sensors - Properties and Applications', Journal of Physics E : Scientific Instruments, 19, 502-515.

6.   McGuire, T.R., and Potter, R.I, 1983, 'Anisotropic Magnetoresistance in Ferro Magnetic Alloys', I.E.E.E. Trans. Magnetics, MAG-11, 1018-1038.

7.   Harmon, L, 1987, 'Automated Tactile Sensing', International Journal Robotics Research, Vol.1, No.2.

8.   Barbera, A.J., et al., 1982, 'Programming of Hierarchical Robot Control System', Proc. 12th Int. Symp. Industrial Robots, Paris.

9.   Hoque, I., 1988, 'Hierarchical Control for Adaptive Gripping', Final Year Project Report, S.E.P., Brunel University.

Chapter 2

3-D robot vision system for industrial applications

P. C. Wong, M. T. Thorn and S. Littlewood

Abstract

This paper descibes a 3-D robot-vision system for industrial applications in terms of cost-effectiveness and efficiency. With the aid of an angled mirror, orthogonal camera planes were analysed through a high-speed vision processor. Calibration was designed to enable flexibility and transportability, and to reduce linear and non-linear errors in the system.

Introduction

Most engineering problems are concerned with finding the most cost-effective and simplest solution. Over the years, many robot-vision techniques have been developed to solve 3-D problems. Most of these such as laser[8], multicameras[2], or stereoscopic analysis[6,10] are still in the trend of current research, and are complex and expensive to implement. This paper proposes a 3-D method which was originated from a two-orthogonal-camera technique. The system was originally proposed to perform chocolate palletising in a confectionery environment. It is a relatively fast and inexpensive method for assembly lines since it uses simple algorithms and an inexpensive sensor : a camera.

Camera and World Modelling

The problem of perspective transformation forms the basis of all robot-vision systems, especially those involving the use of cameras. It concerns the relationship between the 3-D world coordinate system and the camera coordinate system. Figure 1 exhibits the optical projection of a plane in 3-D space onto a camera plane, the two planes being parallel. In a 2-D system, the position of $P(X_w, Y_w)$ is easily obtained if X_c and Y_c are known, and is given by,

$$X_w = \frac{X_c \cdot (\mu - D_w)}{\mu} \quad [1] \quad ; \quad Y_w = \frac{Y_c \cdot (\mu - D_w)}{\mu} \quad [2]$$

where μ is the focal length situated between the two planes, and D_w is the total length. However in the real world, objects are not flat but usually associated with a third dimension, ie., height. In figure 1, the height (depth) is represented by the Z—axis. If a 3-D object is placed under the system shown in figure 1, the values of X_w and Y_w would be affected and their effective values would then be given as

$$X_w = \frac{X_c \cdot (\mu - D_0)}{\mu} \quad [3] \quad ; \quad Y_w = \frac{Y_c \cdot (\mu - D_0)}{\mu} \quad [4]$$

where $D_0 = D_w + Z_w$.

The above system is fine for 2—D applications, since it gives no expression for Z_w in 3-D space. Suppose a second camera is placed on a plane orthogonal to that of the first, as shown in figure 2, to record the y and z axes respectively. Assuming the distance is D_w, that is, same as before, and the camera has the same focal length , the value of Z_w, expressed in a similar form to [3], is

$$Z_w = \frac{Z_c \cdot (\mu - D_x)}{\mu} \quad [5]$$

where $D_x = D_w + X_w$

Solving X_w and Z_w, we would obtain a complete expression for X_w, Y_w and Z_w.

$$X_w = \frac{X_c \cdot (\mu^2 - \mu \cdot D_w - Z_c \cdot \mu + D_w \cdot Z_c)}{(\mu^2 + X_c \cdot Z_c)} \quad [6]$$

$$Y_w = \frac{Y_{c1} \cdot [\mu^3 + 2\mu X_c Z_c - \mu^2 (D_w + Z_c) - 2D_w X_c Z_c + \mu Z_c D_w]}{\mu \cdot (\mu^2 + X_c \cdot Z_c)} \quad [7]$$

$$Z_w = \frac{Z_c \cdot (\mu^2 - \mu \cdot D_w - X_c \cdot \mu + D_w \cdot X_c)}{(\mu^2 + X_c \cdot Z_c)} \quad [8]$$

This model shows that any movement in X_w and Y_w would affect the value of Z_w and vice versa. The neutral point is at the origin where all the values are null; the 'cross-over effect' is greatest at the critical point, ie when an object is placed such that it appears on the camera's edge.

Single Camera Technique

Discrepancies are expected to be found in the lenses and output characteristics of cameras. For this reason, it was decided to substitute an angled mirror for the second camera, to reflect the second plane onto the first camera. This configuration has a number of advantages :

1. Cost saving on hardware, especially on the cost of a camera and the framestore, extra microprocessor or multiplexer that are associated with it.
2. Simplifying the complexity of the calibration process.
3. Removing camera matching errors, output characteristics discrapencies and optical distortion.

However, the immediate question was : Would the above assumptions and expressions still hold ? Figure 3 shows a simplified diagram of the scene. If the mirror is positioned such that the incident ray is parallel to the world plane, then the only variable that needs to be corrected is the distance D_w.

The new distance D_w' for the "second camera " is given as

$$D_w' = D_w. \left[Tan(\theta) + \sqrt{(\tan(\theta)^2 + 1}} \right] \quad [9]$$

But if $\tan(\theta)$ is small, then $D_w' \approx D_w$ within a reasonable tolerance. In practice, θ was about 11° at $\mu=20mm$. That gives an error in D_w' of about 21%.

For the general expression in [1], X_w can be expressed as

$$X_w = X_c.(1 - D_w/\mu) \quad [10]$$

As long as the ratio of $D_w:\mu$ is constant, the value of D_w can be increased ie., physically placing the camera further away from the work field, and hence reducing the error. The image size can be adjusted to be the same as before by using a zoom lens. From equation [8], the error in D' is entirely dependent on θ. One way of reducing the value of θ is to increase the focal length. For example, for 30mm focal length, and $D_w=2m$, $\theta\approx7.1°$, hence giving $D_w'\approx 2.264m$, ie., an error of 13%.

With the aid of a mirror, the problem of image storage was also solved. The framestore, displaying live images from the camera, is shown in figure 4. Effectively, the framestore displays the two orthogonal planes, and obviously the trade off is the reduction in resolution. This was dealt with by using a 512 X 512 pixel square framestore, giving an equivalent of 1.25mm/pixel resolution at a focal length of $\mu=30$ mm.

Calibration

The calibration procedure in a robot-vision system is mainly to deal with :

1. Compensation for camera's aspect ratio.
2. Linear camera-world relationship.
3. Non-linear characteristics.
4. Common and absolute reference points for both coordinate systems.

Here, the use of the centre of the gripper's tip as a reference has made the first and second procedures simpler. During the calibration phase, the robot was driven to a point under the field of view and was moved by a known distance under real-time control in all three directions, corresponding to the X, Y and Z-axes. This gives the required values for the matrix M to relate a point P_W in the real world to a point P_C in the camera plane from

$$\underline{P}_W = M \cdot \underline{P}_C \qquad\qquad [11]$$

where M=A.L and is the combined matrix for matrices A, the aspect ratio, and L , the linear relationship.

This calibration technique therefore includes the linear relationship as well as the aspect ratio of the camera. As soon as the gripper is in sight under the field of view, a reference point is recorded. Further reference points were made relative to this one and thus making the whole system more transportable, ie., for a change in the working environment it would only be necessary to re-register a single reference point.

In any TV-camera vision system there is an accummulated amount of non-linearity due to the lens(eg.,wide angle lens), electromagnetic interference and unevenly distributed sensitivity. The parameters of an object that are worst affected by this non-linear characteristics are the area and perimeter. These parameters are important for recognition and need to be compensated by placing an object in the field of view at different random positions (preferably the corners of the work plane) . Area and perimeter were then computed and used for solving linear simultaneous equations, using a Gaussian elimination method. The results were then used to equalise the curve and to estimate the parameters at any other random positions.

Result

The processing time for image analysis for both planes was about 450ms. However, since the system was designed to analyse the image during a robot (PUMA 560) fetch-cycle, the processing time became comparatively insignificant (typical robot fetch-cycle is about 7 sec, at 50% of the robot's full-speed). Therefore both cycles can be scheduled to be performed in parallel, provided that the robot be out of the field of view during image acquisition.

Objects of height varying from 7 mm to 70 mm were tested and the system worked perfectly. Their average length and breath was 10mm X 35mm. Larger objects can be observed by using a wide angle lens to obtained a wider scope of the work plane. In this experiment, the camera was fitted with a zoom lens and the system was tested with focal lengths varying from 20 to 40mm. The changes were handled by the calibration process without any difficulty, but care needs to be taken to ensure that the assumption in equation [8] still holds.The system needs to be calibrated whenever the environment is changed. The calibration process required about one minute to complete.

The non-linear characteristics were also equalised and being illustrated on figure 5. The overall errors of the equalised area estimations were within ±5% .

Conclusion

This paper described the implementation of a robot-vision system with 3-D capability. It uses the principle of a two-orthogonal-technique, with the second camera substituted by a mirror to reflect the orthogonal plane onto the first camera. This configuration was proved to be cost-effective and simple to calibrate, making the system potentially flexible and transportable. The system was originally designed for cholocate palletisation, and would suit any industrial application which requires pick-and-place on assembly lines.

References

* Reference for papers which exploit mirrors in 3-D applications are found in [1,3].

[1] W Teoh, X D Zhang "An inexpensive stereoscopic Vision system for Robots" IEEE International Conference on Robotics 1984. pages 186-189.

[2] J Y S Luh, J A Klaassen " A real-Time 3-D Multi-camera Vision System" 3rd International Conference on Robot Vision and Sensory Control 1983 (RoViSeC3) pages 400-408.

[3] J Amat, V Llario "A Vision system with 3-D Capabilities". IEEE International Conference on Robotics and Automation. USA March 1985. Pages 2-5.

[4] Neelima Shrikhande, George C Stockman "Feature Matching for Robot Vision". International Symposium on New Direction in Computing. August 1985 USA. pages 103-111.

[5] S.M.Cotter, B G Batchelor "Visual monitoring of palletising and packing". Proceedings of SPIE International Society of Opt. Eng. (USA) Vol.397 pages 240-251. 1983.

[6] Y Kuno, H Numagami, M Ishikawa, H Hoshino, M Kidode "Three dimensional vision techniques for an advanced robot system". IEEE international conference on Robotic and Automation 1985. Pages 11-16.

[7] K S Fu,RC Gonzalez, CSG Lee."Robotics-Control, sensing, vision and Intelligence." 1987 . McGrawHill U.S.A.

[8] James P Simmons "A real time 3-D Vision system for robot guidance" Robotic Engineering, Jan 1986 Vol-8 pages 23-25

[9] H M Morris "Robots see in 3-D". Control Engineering(USA) Jan 1986. Pages 72-75.

[10]Stanley R Sternberg"Three dimensional vision for bin picking" Vision 85 Conference Proceedings (USA) March 1985 pages 2.61-2.67 .

--   --

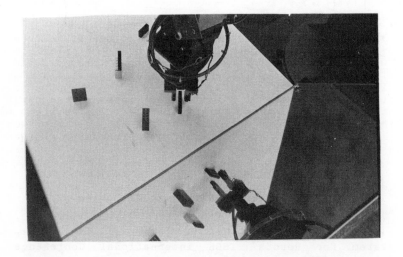

Diagram showing the robot picking up some pseudo objects

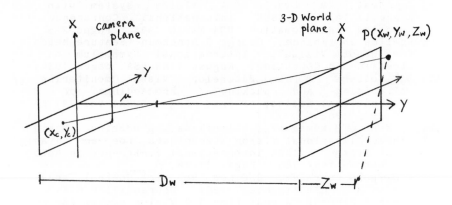

Figure 1. Optical model for 2-D vision system.

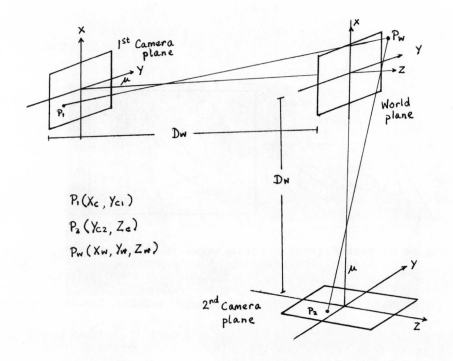

$P_1(X_c, Y_{c1})$

$P_2(Y_{c2}, Z_c)$

$P_w(X_w, Y_w, Z_w)$

Figure 2. Optical model of a 3-D vision system.

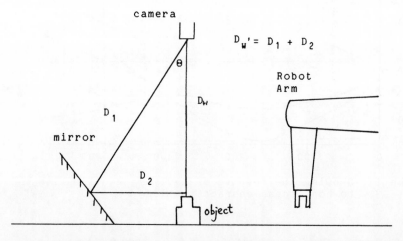

$D_w' = D_1 + D_2$

Figure 3. Simplified diagram showing the acquisition of two orthogonal planes.

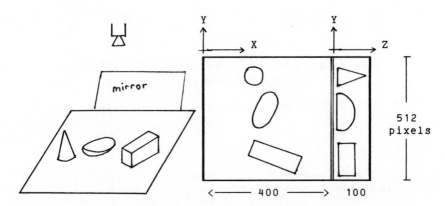

Figure 4. Reconfigured framestore to display the orthogonal
planes. The boundary is programmable.

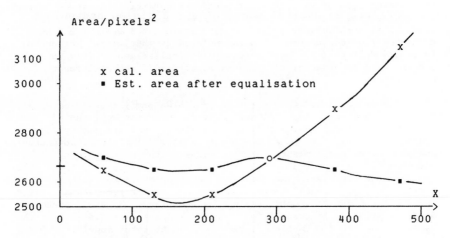

Graph 1 : Showing the results of applying equalisation
to non-linear characteristics.

Chapter 3

An LQG approach to self-tuning control with applications to robotics

S. A. Carr, G. Anderson, M. J. Grimble and J. Ringwood

1. INTRODUCTION

Self-tuning control has been recognised as an effective approach for mechanical manipulator control design due to its ability to cope with the presence of nonlinearities and uncertainties in robot dynamic models. The vast majority of existing self-tuning controllers are based on a linear plant description, the fact that most industrial processes are nonlinear is taken into account by regarding the plant as a sequence of pseudolinear descriptions. Therefore, as the plant operating point changes, the nonlinear plant dynamics are reflected as time varying parameters in the linear plant description. The applicability of this approach has been demonstrated by Koivo and Guo [1] where manipulator joint angular position is the controlled variable. This work is extended in Koivo et al [2] to the case in which the manipulator is controlled directly in the cartesian coordinate system. It is found that convergence of the parameter estimates may not be achieved during the finite time over which the motion takes place. Therefore this approach is particularly suited to repetitive tasks where the last estimates from the previous run can be used as the initial estimates. Lelic and Wellstead [3] have successfully applied generalised pole placement to the control of a 5 axis electrically actuated robot manipulator.

This paper considers control of the joint angular position of a Puma 500 robot arm using a Linear Quadratic Gaussian self-tuning controller. LQG based controllers are widely used since they offer a guarantee of stability (when the plant is known) for open loop unstable and nonminimum phase plants for all values of the cost function parameters. In addition LQG controllers are extremely flexible in the control objectives which may be achieved by appropriate choice of these parameters (for example, integral action for offset removal may easily be introduced via the control weighting function). An ARMAX system model (Section 3) is assumed in the optimal controller design. It can be shown that if the continuous time, nonlinear robot model is linearised and discretised at various operating points, the resulting simplified models are indeed of ARMAX form where the $C(z^{-1})$ polynomial (coloured disturbance) models the constant term resulting from gravity loading.

The aim of this paper is to introduce the reader to the field of self-tuning control and to demonstrate its applicability to robot control by the use of some illustrative examples. Section 2 describes the nonlinear, continuous time model of the PUMA 500 which is used for simulation and analysis. For those unfamiliar with the area, Section 3 introduces the concept of self-tuning control, placing particular emphasis on the process model and parameter estimation scheme employed in this paper. In section 4 the optimal LQG controller is presented in both its explicit and implicit forms. Only the single input/single output solution is presented here but the results may be extended to the multivariable case, Grimble [4]. Section 5 provides examples demonstrating the performance of the self-tuning controller for step changes in the reference angle, Section 6 considers the tracking situation where the reference angle varies sinusoidally. In both cases issues such as closed loop stability, control weighting and conditions for good parameter convergence are considered and these are discussed in greater detail in Section 7. Section 8 concludes the paper with a summary of the ideas which have been presented.

2. ROBOT MODEL

The PUMA 500 industrial robot has six degrees of freedom as shown in Figure 1. The waist, shoulder and elbow joints dictate the end effector position while the wrist determines orientation. The following second order dynamic model of the PUMA 500 (Paul [5]) describes joint angular position in terms of joint torque:

$$T_i = \sum_{j=1}^{3} D_{ij}\ddot{q}_j + Ia_i\ddot{q}_i + \sum_{j=1}^{3} \sum_{k=1}^{3} D_{ijk}\dot{q}_i\dot{q}_k + D_i \qquad (1)$$

where

(i) T_i is the torque at joint i.

(ii) q_i is the angular position of joint i. Likewise $\dot{q}_i$ and $\ddot{q}_i$ represent the joint angular velocity and acceleration respectively.

(iii)   The D_{ii} terms are multiplied by the angular acceleration of the ith joint and as such they are a measure of its effective inertia, the D_{ij} terms represent coupling inertias between the joints. The terms D_{ii} are often overshadowed by the effect of the reflected motor inertia Ia_i, which is frequently large in comparison.

(iv)   Terms of the form D_{ijj} and D_{ijk} represent the centripetal and coriolis torques respectively acting on joint i. A centripetal torque is a torque which acts inwards on any body which rotates or moves along a curved path and D_{ijj} is the centripetal force at joint i due to velocity at joint j. D_{ijk} represents coriolis forces at joint i due to velocities at joints j and k. Coriolis forces arise in cases of motion relative to a moving axis where the motion of the axis produces a change in the direction of the velocity of the mass.

(v)   Finally, D_i represents the gravity loading at joint i.

The coriolis and centripetal torques are important only when the manipulator is moving at high speed. Both the inertial and gravity terms are important in manipulator control as they affect the servo stability and positioning accuracy.

The robot actuator dynamics are now incorporated into the model. Each joint is driven by a permanent magnet D.C. motor, the dynamics of which are described by:

$$V_i = R_i i_i + L_i \frac{di_i}{dt} + k_i^e \frac{d\omega_i}{dt} \qquad (2)$$

$$T_i = k_i^T i_i \qquad (3)$$

$$\omega_i = N_i q_i \qquad (4)$$

where

R_i = Armature resistance

L_i = Armature inductance

N_i = Gear ratio

i_i = Armature current

k_i^e = Electrical time constant

k_i^T = Torque time constant

ω_i = Armature position.

The following third order model may be derived by substituting for T_i in equation 2 using equations 1, 3 and 4.

$$
\begin{aligned}
V_i = {} & \frac{L_i}{k_i^T} \sum_j^3 D_{ij} \dddot{q}_j + \frac{L_i}{k_i^T} Ia_i \dddot{q}_i + \frac{L_i}{k_i^T} \sum_j^3 \sum_k^3 D_{ijk} \ddot{q}_j \dot{q}_k \\
& + \frac{L_i}{k_i^T} \sum_j^3 \sum_k^3 D_{ijk} \dot{q}_j \ddot{q}_k + \frac{R_i}{k_i^T} \sum_j^3 D_{ij} \ddot{q}_j + \frac{R_i}{k_i^T} Ia_i \ddot{q}_i + \frac{L_i}{k_i^T} \sum_j^3 \dot{D}_{ij} \ddot{q}_j \\
& + \frac{L_i}{k_i^T} \sum_j^3 \sum_k^3 \dot{D}_{ijk} \dot{q}_k \dot{q}_j + \frac{R_i}{k_i^T} \sum_j^3 \sum_k^3 D_{ijk} \dot{q}_j \dot{q}_k \\
& + k_i^e N_i \dot{q}_i + \frac{L_i}{k_i^T} \dot{D}_i + \frac{R_i}{k_i^T} D_i \qquad (5)
\end{aligned}
$$

This equation may be written in the general nonlinear state space form as

$$\dot{x} = f(x) + g(x).u \qquad (6)$$

having defined the state vector for the first three joints as

$$x = (x_1 \quad x_2 \quad x_3 \quad x_4 \quad x_5 \quad x_6 \quad x_7 \quad x_8 \quad x_9)^T$$

where

$$x_1 = q_1 \quad ; \quad x_2 = q_2 \quad ; \quad x_3 = q_3$$

$$x_4 = \dot{q}_1 \quad ; \quad x_5 = \dot{q}_2 \quad ; \quad x_6 = \dot{q}_3$$

$$x_7 = \ddot{q}_1 \quad ; \quad x_8 = \ddot{q}_2 \quad ; \quad x_9 = \ddot{q}_3$$

and u is the system input.

Using a classical fourth order Runge Kutta numerical integration of equation 6 a solution for the state vector may be found once the actuator input voltage is specified.   For simulation of the control scheme the motor voltage is calculated from joint position error using an LQG criterion.

3.   SELF-TUNING

An adaptive control system is one in which the controller is automatically adjusted to compensate for unanticipated changes in the process or environment. Adaptive control systems therefore provide a systematic approach for dealing with nonlinearities such as those encountered in robotic systems.

Self-tuners, which estimate model parameters on-line and adjust controller settings accordingly fall into this category.

A typical self-tuning scheme is shown in Fig. 2.   It consists of four main blocks.

(i)   The Process/Process Model

(ii)   The Parameter Estimator

(iii)   Controller Design

(iv)   The Controller

Each block will now be considered in greater detail.

3.1 The Process Model

The ARMAX (AutoRegressive Moving Average with eXogeneous inputs) linear process model is used in this paper to approximate the robot dynamics at the operating point for controller design purposes.   This model describes the plant output in terms of a linear combination of previous plant outputs and delayed control inputs and a coloured noise disturbance (Eqn. 7)

$$y(t) = -a_1 y(t-1) - a_2 y(t-2) - \ldots - a_n y(t-n)$$

$$+b_0 u(t-k) + b_1 u(t-k-1) + \ldots + b_m u(t-k-m)$$

$$+\zeta(t) + c_1 \zeta(t-1) + \ldots + c_\ell \zeta(t-\ell) \qquad (7)$$

where

$y(t)$ = process output at time t

$u(t)$ = control signal at time t

$z(t)$ = white Gaussian noise disturbance

k = process delay.

This can be written more compactly as

$$A(z^{-1})y(t) = B(z^{-1})u(t) + C(z^{-1})\zeta(t)$$

where

$$A(z^{-1}) = 1 + a_1 z^{-1} + \ldots + a_n z^{-n}$$

$$B(z^{-1}) = z^{-k}(b_0 + b_1 z^{-1} + \ldots + b_m z^{-m})$$

$$C(z^{-1}) = 1 + c_1 z^{-1} + \ldots + c_\ell z^{-\ell} \qquad (8)$$

and z^{-1} is the backward shift operator.

3.2   The Parameter Estimator

Identification is the process of constructing a mathematical model (in this case an ARMAX structure is used) of a system from observations and prior knowledge. Identification and parameter estimation have found applications in areas as diverse as engineering, science, economics, medicine, ecology and agriculture.

For the purposes of parameter estimation the ARMAX system description is rewritten in the form

$$y(t) = \Phi^T(t-1).\theta + \varepsilon(t) \qquad (9)$$

where

$$\Phi^T(t-1) = (-y(t-1) - \ldots - y(t-n) : u(t-k) \ldots u(t-k-m)$$

$$\zeta(t-1) \ldots \zeta(t-\ell))$$

$$\theta^T = (-a_1 \ldots -a_n ; b_0 \cdots b_m : c_1 \cdots c_\ell)$$

and

$$\varepsilon(t) = \xi(t)$$

$\Phi^T(t-1)$ is the data vector which contains information about the process up to and including sample time $(t-1)$. θ is the system parameter vector which is to be estimated and $\varepsilon(t)$ represents the estimation error which is assumed to be statistically independant of the inputs and outputs.

A recursive parameter estimation algorithm will, for a specified system output $y(t)$ and data vector $\Phi(t-1)$, find the estimates, θ, of the unknown parameters which minimise a specified loss function $V(\theta)$.

For a quadratic criterion (where the objective is to minimise the squared difference between actual and estimated plant outputs):

$$V(\theta) = 1/2. \; E\left[y(t) - \Phi^T(t-1).\theta \right]^2$$

where $E[.]$ denotes the expectation operator.

Minimisation of the loss function, $V(\theta)$, in conjunction with an ARMAX plant description yields the following Extended Least Squares algorithm.

3.2.1   Extended Least Squares Algorithm (ELS).

$$\theta(t) = \theta(t-1) + L(t)\left[y(t) - \theta(t-1).\Phi(t-1) \right] \tag{10}$$

$$L(t) = \frac{P(t-1)\psi(t-1)}{1 + \psi^T(t-1).P(t-1).\psi(t-1)} \tag{11}$$

$$P(t) = P(t-1) - \frac{P(t-1).\psi(t1)\psi^T(t-1)P(t-1)}{1 + \psi^T(t-1).P(t-1)\psi(t-1)} \tag{12}$$

where

$P(t)$ = Covariance matrix

$L(t)$ = Gain vector

$\psi(t)$ = The gradient vector

$$= \frac{dy(t,\theta)^T}{d\theta} = \frac{d\Phi^T\theta}{d\theta}$$

$= \Phi^T(t-1)$ under the assumption that $\Phi^T(t-1)$ is independant of θ.

3.3   Controller Design

Once the model parameters have been estimated they are used to design the controller. This can be done in two ways.

(i)   **Explicit Algorithms**
        The estimated process parameters are manipulated mathematically to produce the updated set of controller parameters.

(ii)   **Implicit Algorithms**
        The process model is parameterised in terms of the controller parameters in such a way as to update them directly at the identification stage.

The implicit approach gives some advantages with respect to computational speed while the explicit algorithms are more flexible in the different control objectives which can be achieved.

3.4   The Controller

The controller used is a fixed structure controller, the parameters of which are varied by the design stage. The choice of control law depends upon many factors, including,

(i)   Is the process open loop unstable?

(ii)   Is it non-minimum phase?

(ii)   Is there a significant dead time?

(iv)   Is the measured output corrupted by noise?

(v)   Are there constraints on available computational power?

(vi)   Is excessive actuator movement undesirable?

Therefore the choice of control law is highly application dependent. Ultimately a trade-off must be made between improved quality of control and controller simplicity.

For this paper, the control law under consideration is based on an LQG criterion and is discussed in the next section.

4.   LINEAR QUADRATIC GAUSSIAN CONTROL

The fundamental difference between LQG control laws and those which are based on a minimum variance type solution is that LQG controllers are based on the minimisation of a cost function containing an unconditional expectation operator while the expectation is conditional for minimum variance solutions.   Therefore minimum variance control laws are suboptimal with respect to LQG control laws.

The closed loop discrete time system decription used (Fig. 3) is given by the following set of equations:

(i)   **System output equations**

$$y(t) = A^{-1}(z^{-1})(B(z^{-1})u(t) + C(z^{-1})\xi(t))$$

This is the ARMAX plant model which is described in Section 3.1.

(ii)   **Observation process**

$$z_0(t) = y(t) + v(t)$$

where $v(t)$ is an output disturbance of variance R.

(iii)   **Controller input**

$$e_1(t) = r(t) - z_0(t)$$

The controller is fed by the difference between the desired output and observed output.

(iv)   **Reference Generation Process**

$$r(t) = A^{-1}(z^{-1})E(z^{-1})\omega(t)$$

Provision is made for a stochastic reference which is generated by the above subsystem. A deterministic set point may also be included.

(v)   **Tracking error**

$$e(t) = r(t) - y(t).$$

Tracking error is the difference between the desired output and the uncorrupted plant output.

In future discussions the arguments of the polynomials and variables will be omitted for notational simpliciity.

It is assumed that none of these subsystems includes unstable hidden modes (thus unstable and uncontrollable modes are not present).

The following unconditional cost function is to be minimised.

$$J = E\left[Q_1 e^2(t) + R_1 u^2(t)\right] \tag{13}$$

where

$e(t)$ = Error signal (difference between reference and output)

$u(t)$ = Control signal

Q_1 = Error weighting

R_1 = Control weighting

The signals r, z and ω are assumed to be stationary, zero mean sequences of uncorrelated random variables which have variances given by R, Q_2 and

Q_3 respectively. The generalised spectral factors satisfy:

$$Y\bar{Y} = (EQ_3\bar{E} + CQ_2\bar{C} + AR\bar{A})/(A\bar{A}) \tag{14}$$

$$\bar{Y}_1 Y_1 = (\bar{B}Q_1 B + \bar{A}R_1 A)/(\bar{A}A) \tag{15}$$

where

$$\bar{X}(z^{-1}) = X(z)$$

is called the adjoint of $X(z^{-1})$.

The strictly Hurwitz (stable inverse) polynomials in z^{-1}, D and D_1 may be defined as

$$Y\bar{Y} = D\bar{D}/(A\bar{A}) \Rightarrow Y = D/A \tag{16}$$

$$\bar{Y}_1 Y_1 = \bar{D}_1 D_1 / (\bar{A}A) \Rightarrow Y_1 = D_1/A \tag{17}$$

4.1 Explicit LQG

The solution for the optimal explicit LQG controller may now be presented, a more detailed derivation may be found in Grimble [6]. The optimal controller transfer function is:

$$C_o = G_o/H_o \tag{18}$$

where G_o and H_o are polynomials in z^{-1}. The following coupled diophantine equations in terms of the unknown polynomials G, H and F, provide the unique particular solution G_o, H_o with minimal degree with respect to F:

$$\bar{D}_1 z^{-g} G + FA = \bar{B} z^{-g} Q_1 D \tag{19}$$

$$\bar{D}_1 z^{-g} H - FB = \bar{A} z^{-g} R_1 D \tag{20}$$

where $g \triangleq \max(n_{d1}, n_b, n_a)$. These equations can be combined to obtain the implied equation:

$$AH + BG = D_1 D \tag{21}$$

The solution of the implied diophantine equation (21) with $n_h < n_b$ and $n_g < n_a$ and A,B coprime is unique, as may be verified from Theorem 4, Ježek [7].

It can be seen that equation (21) is in fact the closed loop characteristic polynomial and as DD_1 can be shown to be strictly Hurwitz the stability of the system is guaranteed, regardless of the control weighting even for open loop unstable or nonminimum phase plants.

4.2 Implicit LQG

Implicit self-tuning controllers sometimes have advantages for implementation due to their direct means of calculating the controller parameters. An implicit LQG control law may be derived assuming that a unique solution to (21) exists (A and B are coprime). The following innovations plant description is used:

$$A(z^{-1})e_1 = D(z^{-1})\varepsilon - B(z^{-1})u \tag{22}$$

where $A(z^{-1})$, $D(z^{-1})$, e and u are as previously defined and ε is a unit variance white noise signal. This closed loop model can be shown to be equivalent to that of Fig. 3. Combining equations (21) and (22) yields

$$Ae_1 = (AH_o + BG_o)\varepsilon/D_1 - Bu$$

After some manipulation this can be shown to be equivalent to

$$\phi(t) = H_o \varepsilon + B/D(G_o e_1 - H_o u) \tag{23}$$

where $\phi(t) = D_1 e_1$

This represents the desired implicit model from which G_o and H_o may be estimated. However if $n_{ho} > k-1$ then the residual $H_o \varepsilon(t)$ is correlated with the regressors $e_1(t-k)$, $e_1(t-k-1)$,..$u(t-k)$,$u(t-k-1)$ in (23). Thus write $H_o = H_{o1} + H_{o2}$ where H_{o1} includes all the terms with powers of z^{-1} up to $z^{-(k-1)}$ to obtain the least squares predictor:

$$\hat{\phi}(t|t-k) = H_{o2}\varepsilon(t) + B/D(G_o e_1(t) - H_o u(t)) \tag{24}$$

and the prediction error

$$\tilde{\phi}(t|t-k) = H_{o1}\varepsilon(t) \tag{25}$$

where $\hat{\phi}(t|t-k)$ denotes that the value of $\hat{\phi}$ at time t is based only on data up to time t-k. As the optimal control signal $u^o(t)$ is chosen to set the final term in (24) to zero, the prediction equation can be simplified as:

$$\hat{\phi}(t|t-k) = H_{o2}\varepsilon(t) + B(G_o e_1(t) - H_o u(t)) \tag{26}$$

using the argument employed by Clarke and Gawthrop [8].

If the polynomials A,B and D and the innovations signal ε are estimated using ELS parameter estimation and equation (22) and the stable spectral factor D_1 is calculated using (17) then by defining $e_b = Be_1$ and $u_b = Bu$ the controller polynomials G_o and H_o can be identified using

$$\hat{\phi}(t|t-k) = G_o e_b - H_o u_b + H_{o2}\varepsilon$$

and Extended Least Squares Identification

5.   SET POINT CONTROL

This section evaluates the regulatory performance of the self-tuning controller. The response of the robot arm to the motor actuation voltage (controller output) is simulated using the third order nonlinear state space model of equation 6. The Runge-Kutta integration time used is 1 millisecond. The controller sampling time was chosen, based on open loop step response curves, to be 0.1 seconds. At each control sampling interval the data vector for parameter estimation (Section 3.2) is updated, the coefficients of the A,B and C polynomials are re-evaluated using the new data vector, the explicit LQG controller is designed based on the updated estimates and the new actuation signal is applied to the robot model. All simulations consider control of the angular position of joint three, with joints one and two locked at zero radians. Linearisation and discretisation of the robot model of equation 5 at various operating points shows that the linearised system is best described by an ARMAX model where the $A(z^{-1})$ and $B(z^{-1})$ polynomials are third order and the $C(z^{-1})$ polynomial which

characterises the gravity loading term is second order.   The
polynomials $G(z^{-1})$, $F(z^{-1})$ and $H(z^{-1})$ are chosen to be second order to
balance the powers of z^{-1} on both sides of the diophantine equations
(19) and (20).

Example 1

Figure 4a shows the angular position of joint three when the
reference angle is zero radians (vertical joint).   The initial $A(z^{-1})$,
$B(z^{-1})$ and $C(z^{-1})$ parameter estimates (Figures 4b - d) were chosen
based on the final estimates from previous trials.   The initial values
of the diagonal elements of the covariance matrix, $P(t)$, where chosen to
be 1000 to promote rapid convergence since the magnitude of step
increments/decrements in the parameter estimates at each iteration is
directly dependent on the magnitude of the elements of $P(t)$.

The open loop system transfer function, $B(z^{-1})/A(z^{-1})$, converges
to

$$B(z)/A(z) = \frac{0.02(z + 1.025)(z-0.975)}{(z-.6298)(z-1.27)(z-1)}$$

which is both unstable and nonminimum phase.   The closed loop transfer
function, $G(z)$, is:

$$G(z) = \frac{0.64z(z+1.025)(z-0.975)\left[(z-0.662)^2+0.1842^2\right]}{(z+0.97)\left[(z-0.69)^2+0.272^2\right](z-0.442)(z-0.974)} \qquad (27)$$

Therefore it can be seen that the LQG self-tuning controller has
stabilised the system.

The joint angle is initially disturbed from the reference position
while the parameters tune in, reaching a maximum angle of -1.65
radians.

Example 2

The initial output disturbance in the previous example is clearly
undesirable.   This example shows that a dramatic improvement in
performance may be obtained by using a fixed LQG controller during the
initial tuning-in period (Figure 5).   The fixed controller is designed
based on the estimates of the polynomials $A(z^{-1})$, $B(z^{-1})$ and $C(z^{-1})$ of
Figure 4.   The maximum deviation of the output from the reference is
reduced to approximately 0.1 radians, less than 1% of that in the
previous example.

Example 3

Figure 6 presents the closed loop response of the system for
different step inputs.   The converged values of the $A(z^{-1})$, $B(z^{-1})$ and
$C(z^{-1})$ polynomials at various operating points, including those of
figure 6, are given in Table 1. Analysis of the $A(z^{-1})$ and $B(z^{-1})$
polynomials shows that at each operating point the open loop system is
unstable and nonminimum phase, whilst the closed loop system is stable.

Each response is characterised by a time constant of approximately 1.5 seconds, an initial overshoot of between 10% and 20% of the set point reference and a settling time of approximately 3 seconds.   In each case there is a small steady state offset (less than 0.01 radians) due to the fact that the nonlinear system is represented, for controller design, by a linear model.

6.   PATH TRACKING

This example considers the more realistic situation where the required joint angular position is not stationary but varies with time.

Example 4

For this simulation a sinusoidal reference angle with a period of 10 seconds, a peak to peak amplitude of 1.6 radians and a mean value of 1.0 radians was applied to the system. Figure 7a shows that after an initial overshoot during the tuning in period (of approximately 3 seconds) the actual angle of joint three tracks the specified angle closely. As the link moves through its nonlinear region of operation the parameter estimator revises, at every sample instant, the linear plant description which is used for controller design. Figures 7b to 7d show the variation with time of the $A(z^{-1})$, $B(z^{-1})$ and $C(z^{-1})$ parameters respectively. As expected from analysis of the robot model, the A parameters remain constant during the simulation. There is a small variation in the B parameters, particularly in the b_2 coefficient. However, as expected, the most noticeable variation occurs in the coefficients of the $C(z^{-1})$ polynomial, which models the gravitational disturbance term. This is because, under the operating conditions of this example the nonlinear gravitational torque predominates.

7. DISCUSSION

The following factors have been considered in the implementation of the self-tuning control scheme presented in this paper.

7.1 Parameter Tracking

The covariance matrix, $P(t)$, is a positive definite measure of the estimation error — therefore the magnitude of it elements tends to decrease with time.   The magnitude of the step change in θ, the parameter vector, at each iteration is directly dependent on the magnitude of the elements of $P(t)$. To maintain the sensitivity of the algorithm and allow for parameter tracking some modifications must be made to the algorithm of equations (10) to (12) in order that the elements of $P(t)$ are prevented from becoming too small. One technique which is commonly used is to include an exponential weighting factor in the performance index as follows:

$$V(\theta) = \sum_{i=1}^{t} \lambda^{t-i}\left[y(i) - \Phi^{T}(i-1).\theta(i) \right]^2$$

where

$$0.0 < \lambda \leqslant 1.0 \tag{28}$$

for $\lambda = 1.0$ all data are weighted equally. For $0 < \lambda < 1.0$ more weight is placed on recent measurements than on older measurements. This revised performance index results in the following least squares algorithm.

$$\theta(t) = \theta(t-1) + L(t)[y(t) - \theta^T(t-1)\Phi(t)] \tag{29}$$

$$L(t) = \frac{P(t-1).\psi(t)}{\lambda + \psi^T(t).P(t-1).\psi(t)} \tag{30}$$

$$P(t) = 1/\lambda\left[P(t-1) - \frac{P(t-1).\psi(t).\psi^T(t).P(t-1)}{\lambda + \psi^T(t).P(t-1).\psi(t)}\right] \tag{31}$$

7.2 Initial Parameter Estimates

Estimates of the data vector, θ, must be supplied to initiate the algorithm of equations (29) to (31). If no prior knowledge of the system is available then these are chosen arbitrarily. If the elements of the covariance matrix, $P(t)$, are large (of the order of 10^5) and the forgetting factor, λ, is less than 1.0 (a value of 0.95 gives good results) then the elements of the gain vector, $L(t)$, are large and rapid convergence is achieved. In order to ensure excitation of all of the modes of the system a stochastic, persistently exciting reference (Norton [9]) may be applied to the controller initially. It is advisable to use a fixed controller in parallel with process identification during the tuning in period, after which control is transferred to the self-tuner. After the first trial improved parameter estimates are available. These should be used as the initial estimates for the next task. For set point control the initial elements of the covariance matrix should be reduced (to of the order of 10) and the forgetting factor should be increased (a value of 0.98 was used in examples 1 to 3) to decrease the sensitivity of the algorithm.

However, if, as in the vast majority of robotics applications, parameter tracking is required (example 4), these values should be approximately 100 and 0.97 respectively to maintain sensitivity to parameter changes while at the same time rejecting measurement disturbances.

7.3 Offset Removal

From classical control theory it can be shown that offset removal may be achieved, for linear systems, by

(i)   increasing controller gain

(ii)   including integral action in the controller.

The open loop process transfer function of example 1 shows that the numerator is multiplied by a factor of 0.02 relative to the denominator. Therefore a fixed gain of 50 was included in the controller transfer function. This had the effect of signficantly decreasing the steady state offset, since it is known that steady state offset is inversely proportional to controller gain for a step reference input. However, if the gain is increased further, oscillations are introduced in the output as the closed loop poles move along the root locus towards the zeros near the unit circle of (27).

Integral action removes steady state offset for linear systems. However, the system under consideration is highly nonlinear and the inclusion of an integrator is equivalent to placing a controller pole on the unit circle. It was found that this results in an oscillatory response, even when the controller gain is reduced.

The steady state offset of examples 1 to 3 is a result of representing a highly nonlinear system by an approximate linear model. However the magnitude of this offset may be reduced, using a high gain controller, to within tolerable limits for many applications. The use of a virtual reference is also suggested.

7.4 Control Weighting

The control weighting parameter, R_1, determines the relative importance which is to be placed on the penalisation of the control signal by the cost function. It follows that a high value of R_1 leads to less control variation and a more highly 'damped' output while a low relative value of R_1 leads to a lower variance of tracking error and smaller offset. For this reason the ratio of $Q_1 : R_1$ was chosen to be quite high at 1:0.01. It is also possible to use dynamic (frequency dependent) weights in the solution of the optimal controller to allow the error and control signals to be penalised differently in different frequency ranges.

8. CONCLUSIONS

It has been shown with the aid of some illustrative examples that LQG self-tuning control can successfully be used to stabilise and control a mechanical manipulator. For simulation purposes the robot dynamics are described by a set of third order, cross-coupled, nonlinear equations. The self-tuning controller represents these by a linear ARMAX model which is updated at every sample interval. The single joint control examples presented in this paper are simple but demonstrative. They may easily be extended to consider multiple joint control and end effector position control in cartesian coordinates.

9. REFERENCES

1. Koivo, A.J. and Guo, T., 1983, IEEE Trans. Aut. Control, Vol AC-28. 162-171.

2. Koivo, A.J., Kunkel, R., and Guo, T., 1983, 'Adaptive Manipulator Control in Cartesian Coordinate System', Proceeding of the 7th International Computer Software and Applications Conference of the IEEE Computer Society, Chicago, Illinois.

3. Lelic, M.A., and Wellstead, P.E., 1986, 'A Generalised Pole Placement Self-Tuning Controller – An Application to Robot Manipulator Control', Report No. 658, Control Systems Centre, UMIST.

4. Grimble, M.J., 1986, <u>Int. J. Systems Sci.</u>, Vol. 17, No. 4, 543–557.

5. Paul, R.P., 1981, 'Robot Manipulators : Mathematics, Programming and Control', The MIT Press, Cambridge, Massachusetts.

6. Grimble, M.J., 1984, <u>Automatica</u>, Vol. 20, 5, 661–669.

7. Jezek, J., 1982, <u>Kybernetika, Vol. 18, 505–516.</u>

8. Clarke, D.W., and Gawthrop, P.J., 1975, <u>Proc. IEE</u>, Vol. 122, 929–934.

9. Norton, J.P., 1986, 'An Introduction to Identification', Academic Press, London, England.

Table 1   The estimated plant parameters at various operating points

Angle (Radians)	a_1	a_2	a_3	b_o	b_1	b_2	c_1	c_2
-0.50	-2.8	2.6	-0.8	0.0205	0.003	-0.0185	0.95	0.15
0.00	-2.9	2.7	-0.9	0.0200	0.001	-0.0200	0.94	0.12
0.30	-3.0	2.6	-1.0	0.0200	0.002	-0.0190	0.61	0.39
1.57	-2.9	2.9	-0.9	0.0200	0.001	-0.0190	0.40	0.47
2.00	-3.0	2.9	-1.0	0.0235	0.002	-0.0210	0.50	0.56
3.00	-2.8	2.4	-0.8	0.0280	0.001	-0.0180	0.20	0.40

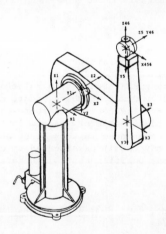

Fig. 1   The PUMA 500 manipulator.

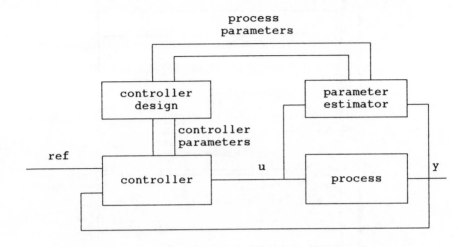

Fig.2   A typical self-tuning control scheme

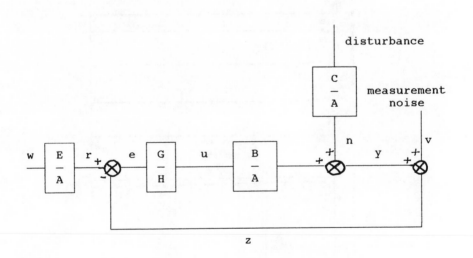

Fig.3   The closed loop system

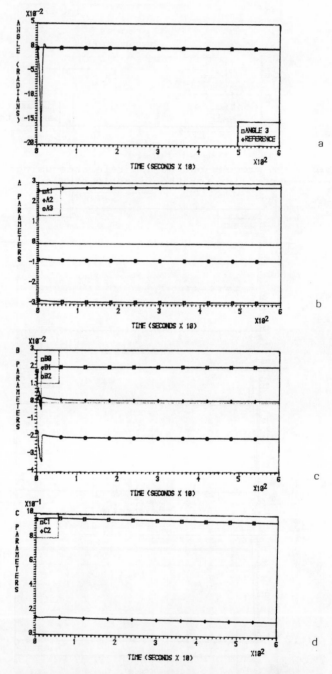

Fig. 4   a) The process output and reference b)The A
polynomial parameters c)The B polynomial
parameters d)The C polynomial parameters.

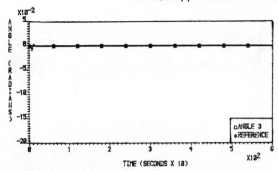

Fig. 5 The process output and reference using a fixed
controller for the first 10 samples.

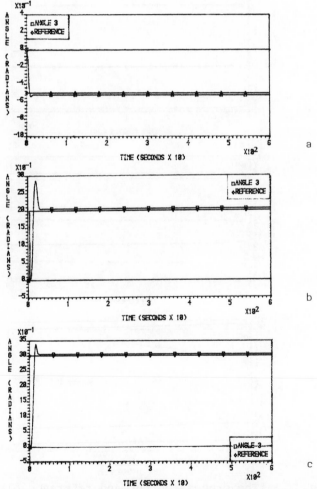

Fig. 6 The process output for a) -0.5 radian set point
b) 2.0 radian set point c) 3.0 radian set point.

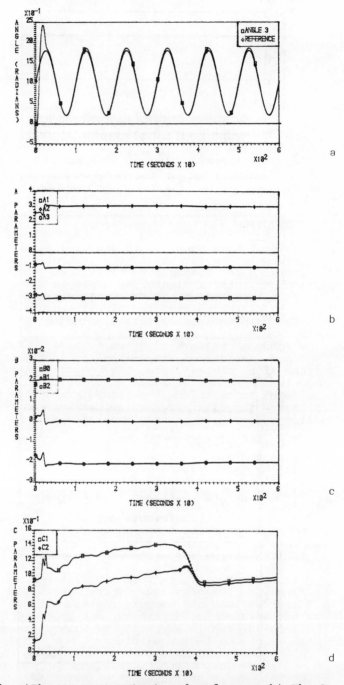

Fig. 7   a)The process output and reference b) The A
polynomial parameters c)The B polynomial
parameters d) The C polynomial parameters.

Chapter 4

Adaptive control algorithms for intelligent robot manipulators

M. Farsi, K. J. Zachariah and K. Warwick

1.   INTRODUCTION

The aim of this paper is to describe adaptive
controllers suitable for use in the control of robotic
manipulators. Upper Diagonal Factorization (UDF) or a
simplified parameter estimator is employed within the self-
tuning algorithms to estimate the parameters contained in
CARMA models of the joints. The simplified estimator used
reduces the computational effort considerably.

The robot under investigation is made by Kuka
(Fig. 1) and consists of six 'links' attached serially to
each other at revolute joints. The coordinate frames
attached to the manipulator are also indicated in the
figure. Joint information is obtained via an optical
encoder measuring relative angular displacement and a
tachometer measuring relative angular speed between
adjacent links.

The manipulator is intended to interact with
objects in the three dimensional space surrounding it,
therefore, it is necessary to obtain the position and
orientation of the end effector in cartesian space. This
information, in this case, in the form of binary data is
made available to a transformation module of the robot
via the vision sensor machine.

2.   DYNAMIC MODEL OF THE KUKA ROBOT

In order to be able to test a variety of control
schemes and also the behaviour of the controlled robot
under different load and speed conditions it is necessary
 to operate on a comprehensive dynamic model. The dynamic
model can be derived in a variety of ways such as the
Newton-Euler or the Lagrange-Euler methods, Ranky and Ho
(1). The approach used here is based on the use of
Lagrangian mechanics, where expressions for the kinetic
and potential energy of the robot structure are used to
obtain relationship between the input (torque) and output
variables. The equations for the first three links of the
Kuka robot are as follows:

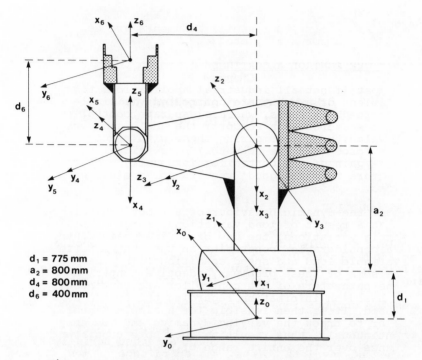

Fig. 1.1   Schematic diagram of the Kuka robot
            manipulator

$$t_1 = D_{11}\ddot{\theta}_1 + D_{112}\dot{\theta}_1\dot{\theta}_2 + D_{113}\dot{\theta}_1\dot{\theta}_3 \quad\ldots\ldots\ldots(2.1)$$

$$t_2 = D_{22}\ddot{\theta}_2 + D_{23}\ddot{\theta}_3 + D_{211}\dot{\theta}_1{}^2 + D_{233}\dot{\theta}_3{}^2$$
$$+ D_{223}\dot{\theta}_2\dot{\theta}_3 + D_2 \quad\ldots(2.2)$$

$$t_3 = D_{33}\ddot{\theta}_3 + D_{32}\ddot{\theta}_2 + D_{311}\dot{\theta}_1{}^2 + D_{322}\dot{\theta}_2{}^2 + D_3 \quad(2.3)$$

where the t's are the joint torque and θ's are
angular displacements. A detailed derivation of the
model is available in (2).

A discrete time (CARMA) model representing each of
the equations (2.1-2.3) may be defined by

$$A(z^{-1})y(t) = z^{-k}B(z^{-1})u(t) + C(z^{-1})e(t) \quad\ldots(2.4)$$

where y(t) and u(t) are the sampled values of joint angle
and motor power, respectively, at time t. e(t) is
considered to be an uncorrelated 'white noise' sequence of
zero mean and finite variance. k $\geqslant$ 1 is an integer time
delay corresponding to the system time lag in terms of the
number of sample intervals.

A, B and C are polynomials in the backward shift

operator (z^{-1}) with the appropriate dimensions. In the
interests of brevity, polynomial arguments are henceforth
omitted, i.e. A for $A(z^{-1})$ etc.

3.   ADAPTIVE CONTROL ALGORITHMS

In most if not all industrial applications today, the
link positions of a robot are controlled via simple fixed
parameter feedback loops, where a separate controller at
each joint measures (via sensors) the angle and perhaps
angular velocity and, comparing these with required values,
applies a force (via the actuator) proportional to the
error. The gain of the controller is usually pre-set by the
manufacturers to give 'good' results for nominal working
conditions.

Since the dynamic behaviour of a robot manipulator
changes as a result of load variations, large displacements,
and inertia variations, the use of a fixed-gain controller
places a considerable constraint on the system performance
under variable load and speed conditions. In order to
overcome some of these problems, adaptive control schemes in
which the parameters of the controller are continuously
updated based on information regarding the system dynamics
can be used. It is then feasible for a simple linear
controller to track the varying dynamics and compensate for
them continuously, thus resulting in a more consistent
performance over the entire range of operating range.

General research into the field of adaptive control
has been progressing for several decades Jacobs (3), but
more recently and mostly due to the advances in computer
technology, its application to robotic manipulators has
also been increasingly investigated Koivo (4), Farsi et al
(5). The following control strategies are investigated for
this application.

3.1   Control Strategy 1, Simple Pole Assignment

This strategy is formulated and achieved by considering
the following control law, Wellstead and Sanoff (6):

$$Fu(t) = Gy(t) + Rr(t) \quad \dots\dots\dots\dots\dots\dots\dots\dots (3.1.1)$$

where F,G, and R are polynomials in z^{-1}, and u(t), y(t),
and r(t) are the input, output and set-point of the plant
respectively.

This control law leads to the following closed loop
equation:

$$y(t) = \frac{CFe(t) + z^{-k}BRr(t)}{AF - z^{-k}BG} \quad \dots\dots\dots\dots\dots (3.1.2)$$

The denominator $(AF - z^{-k}BG)$ is the closed loop characteristic polynomial and defines system response which may be set according to a predefined pole polynomial, T, using the identity:

$$AF - z^{-k}BG = T \quad \dots\dots\dots\dots\dots\dots\dots\dots(3.1.3)$$

The coefficients of polynomials F and G are obtained by solving the identity (3.1.3) whilst R is calculated for offset-free set-point following in the steady state.

3.2   Control Strategy 2, PID Pole Assignment

Due to the immense popularity of the PID controller in industry, it is considered useful to investigate an adaptive form of this type of controller. Here, a special case of the pole assignment strategy is used to achieve the PID control law.

The general structure of the controller is given by (3.1.1) with the polynomials F,G and R defined as:

$$
\begin{aligned}
F &= (1 - z^{-1})(1 + az^{-1}) \\
G &= g_0 + g_1 z^{-1} + g_2 z^{-2} \quad \dots\dots\dots\dots\dots(3.2.1) \\
R &= -(g_0 + g_1 + g_2)
\end{aligned}
$$

where 'a' is an extra parameter used to improve the performance in industrial PID controllers, Wittenmark [7]. Equations (3.2.1) when substituted in (3.1.1) lead to a PID controller with its integral action on the deviation of the output from the set-point and the proportional and the derivative actions on the output.

3.3   Control Strategy 3, General Predictive Control

The general predictive controller, GPC, Clarke [8], is a robust and versatile 'general purpose' algorithm for the stable control of the majority of 'real' processes. The use of a type of plant model which enables the controller to possess inherent integral action, and the facility of selecting a horizon beyond which control changes are given infinite weights in the cost function are some of the key features of GPC. The use of a future set-point sequence for improved performance also makes this algorithm a good candidate for the control of robotic manipulators, since future reference points on the path trajectory are always computed in advance.

The GPC algorithm uses a plant model where the disturbance is realistically modelled to represent real-life situations like random steps occuring at random times and Brownian motion. This leads to the CARIMA model which is given by:

$$A\ y(t) = B\ u(t-1) + n(t)/\Delta \quad \ldots\ldots\ldots\ldots\ldots(3.3.1)$$

where Δ is the differencing operator $(1-z^{-1})$, and $n(t)$ is an uncorrelated random sequence.

The GPC law is derived by minimizing a cost function, J, which is defined as:

$$J = \sum_{j=N1}^{Ny} [y(t+j)-w(t+j)]^2 + \sum_{j=1}^{Nu} \lambda\ [\Delta u(t+j-1)]^2 \quad .(3.3.2)$$

in which w(t+j) is the future reference set-point sequence, k is a control weighting factor, N1 and Ny are the minimum and maximum output prediction horizons and Nu is the control horizon. Minimizing (3.3.2) with respect to Du gives the following control law:

$$Du = (G^T G + \lambda I)^{-1} G^T\ (w-s) \quad \ldots\ldots\ldots\ldots\ldots(3.3.3)$$

where G is a NyxNu matrix, w and s are Ny vectors containing future set-point sequence and components of future outputs which are known at time t respectively. Equation (3.3.3) can be rewritten as:

$$u(t) = u(t-1) + g\ (w-s) \quad \ldots\ldots\ldots\ldots\ldots(3.3.4)$$

in which g is the first row of the NuxNy matrix, $(G^T G + \lambda I)^{-1}\ G^T$.

GPC is a receding horizon control law in which a suggested sequence of future control actions are generated from a sequence of future reference points and output prediction. The future control sequence is recalculated at each sample period and the first element of the sequence i.e. u(t) is applied. The use of a control horizon, Nu, is a useful 'tuning knob' in varying the liveliness of the controller and is a deciding factor in reducing the computational burden of the controller.

4. SIMULATION STUDIES

The performance of the three types of self-tuning controllers described in the previous section when applied to the 3 links of the Kuka manipulator model was analysed using a computer simulation program. Some sample results using the control algorithms investigated are presented.

The simulation studies performed made use of the numerically robust UDF method, Bierman (9), to estimate the parameters of the system. A simplified estimator, Farsi et al (10) which is computationally efficient was also tried out during simulation studies with the GPC algorithm.

(Fig. 4.1 - 4.5) illustrate the response of joint two of the manipulator when moved from 0° to 18° and back again, using different controllers. (Fig. 4.1) is the response when a fixed PD controller is employed. This controller is of the same form as that used in practice. It is noticed that the actual response is lagging the reference point considerably. (Fig. 4.2 -4.5) are responses obtained when using the adaptive control strategies. It is observed that the response using GPC strategy is far superior to the others. The simplified estimator used with GPC also produces acceptable results, with far less computational effort.

5.   CONCLUSION

Results indicate that the adaptive controllers perform satisfactorily when used with robotic manipulators. It is observed that the GPC strategy performs better than others due to the availability of extra 'tuning knobs' as well as the use of the future set-point sequence.

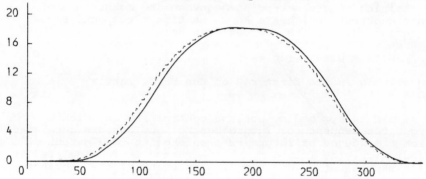

Fig. 4.1   Response of a fixed-gain PD controller

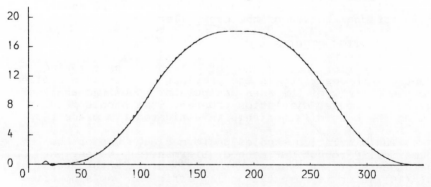

Fig. 4.2   Response of GPC with UDF estimator

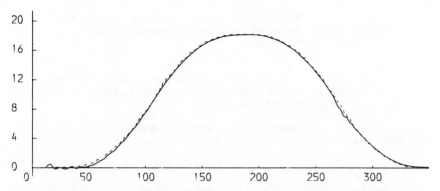

Fig. 4.3   Response of GPC with simplified estimator

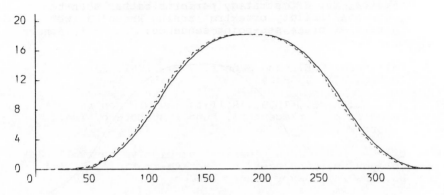

Fig. 4.4   Response of PA controller with UDF estimator

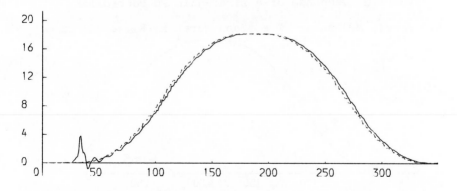

Fig. 4.5   Response of PID controller with UDF estimator

6 REFERENCES

1.   Ranky, P.G. and Ho, C.Y., 1985, 'Robot modelling -
      Control and applications with software', IFS
      Publications Ltd., New York, USA.

2.   Control Group, 1987, 'Report on the preliminary
      dynamic modelling of the Kuka robot', Department
      of Electrical and Electronic Engineering, University
      of Newcastle Upon Tyne, UK.

3   Jacobs, O.L.R., 1961, J. Elect., and Cont., 10, 4,
      311-322.

4.   Koivo, A.J. and Guo, T.H.,1983, IEEE Trans. AC, AC-28,
      162-170.

5.   Farsi, M., Finch, J.W., Warwick, K., Tzafestas, S.G.,
      Stassinopoulos, G., 1986, 'Simplified PID Self-tuner
      Controller for Robot Manipulators', The 25th IEEE
      Conference Proceeding, 'Decision and Control', Athens,
      Greece.

6.   Wellstead, P.E., and Sanoff, S.P., 1981, Int. J.
      Control, 34, 433-455.

7.   Wittenmark, B., 1979, 'Self-tuning PID controllers
      based on pole placement', Lund Institute of Tch.,
      report No. TFRT-7179.

8.   Clarke, D.W., 1986, 'General predictive control', IEE
      Colloquium, 'advances in adaptive control', Digest
      No. 1987/22, Savoy Place, London, UK.

9.   Bierman, G.J., 1977, 'Factorization Methods for
      Discrete Sequential Estimation' Academic Press, New
      York, USA.

10.   Farsi, M., Karam, Z.K., and Warwick, K., 1984, Elect.
      Letters, Vol. 20, No. 22, 913-915.

Chapter 5

Model reference adaptive control for flexible manipulators

J. Z. Sasiadek and R. Srinivasan

1.   INTRODUCTION

In the last forty years, industry has seen an ever-increasing trend towards automation. The last decade has seen robots and manipulators beginning to contribute significantly towards this automatization. In fact, robots would comprise an integral part of the "factory of the future", which is beyond the conception stage and well into its experimentation stage. Such a fully automated factory would operate with minimal human intervention and, could be remotely supervised. But before robots can be implemented on such a large scale, the problem of their excessive weight needs to be urgently addressed. Today's industrial robots usually have several degrees of freedom (4 to 6), and are supported by sophisticated controllers and several sensors. The end-effector, or the end of the last link, which is usually of paramount interest to us, is controlled by controlling the torques and forces applied at the various joints. This method of joint control translates into the control of the end-effector, only when the link kinematics remain constant. This would mean that the shape and geometry of the structural links of the robot need to remain unchanged, even when the robot carries a heavy payload and executes a fast movement. The positional accuracy of a robot can be degraded considerably, if this exacting condition on rigidity is not maintained. But in order to construct the links of the robot for rigidity, one must pay the price in terms of weight. Heavy links require large actuators to move them. Large actuators will in turn consume more power. This will also affect the payload capacity as well as the speed of operation adversely.

All these problems, which stem from the excessive weight of the links, can be overcome by designing a light-weight flexible manipulator. Effective control of such a flexible manipulator would render it invaluable in space and undersea exploration, as well as in remote or hazardous applications.

This paper describes a method based on model-reference adaptive control (MRAC) to control a light-weight, flexible manipulator. In the past, various types of linear control methods, including feedback control, have been applied to achieve positional accuracy of a flexible manipulator. These methods have relied on linearized models of the non-linear dynamics of a manipulator, to achieve effective control. The quality of such control was low. That can be improved by implementation of adaptive control systems. The MRAC scheme proposed here, as its name suggests, consists of the controlled plant, a reference model, a comparator, an adaptation algorithm and a PID controller. The adaptation algorithm is also responsible for non-linearity compensation and decoupling control. The method takes into account the deflection of the links during motion, and automatically compensates for it.

The results of experiments show conclusively that this method is indeed capable of achieving precise positioning control of flexible manipulators.

2.   PROBLEM FORMULATION

The problem addressed here was to develop an efficient method of controlling a flexible manipulator. The objective was to drive the end-effector of the flexible manipulator to follow a given trajectory.

The general features of the MRAC method used here have been described in detail by Landau (1). Applications of that method to robotic control have been described by Sasiadek and Srinivasan (2,3) as well as by Horowitz and Tomizuka (4).

The MRAC method was initially tested on a two-link manipulator model (Fig. 1). This model had both links of the same length (1 m.), and uniform circular cross-section. The following assumptions were made to simplify the analysis:
1.   the links are constrained to move within a plane – the vertical one;
2.   the bending of the links is within the vertical plane only;
3.   actuator dynamics are neglected;
4.   the joints are treated as frictionless.

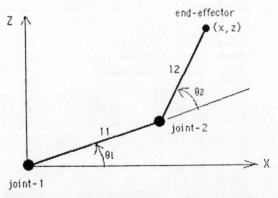

Fig. 1 Flexible Manipulator Configuration

3.   MODEL REFERENCE ADAPTIVE CONTROL

MRAC, as mentioned earlier, is used to control the flexible manipulator. The manipulator control system has been shown in the form of a block diagram (Fig. 2).

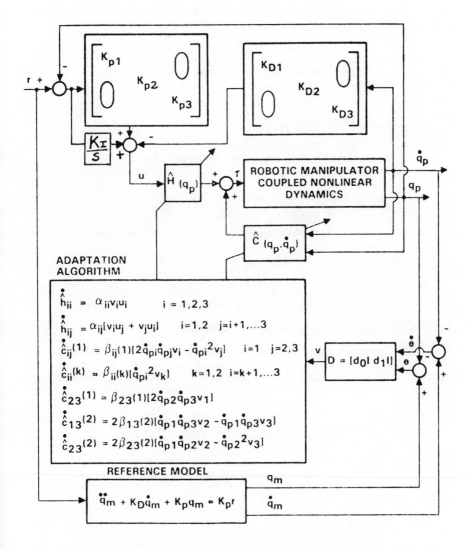

Fig. 2   MRAC Structure for Manipulator Control

A mathematically defined reference model is operated in parallel with the manipulator. Both the model and the manipulator are essentially controlled by a PID controller with position and velocity feedback. The reference model, which is considered to be an ideal system, is represented by a double integrator. This enables the generalized acceleration vector u to be fed into the reference model and obtain the generalized position $\hat{q}_p$ as its output. This ideal position vector $\hat{q}_p$ is compared with that of the manipulator q_p, and the difference between the two yields the position error vector e. A similar procedure with the velocity vectors yields $\dot{e}$. As in any model reference adaptive control system, these errors are then used by the adaptation algorithm to modify the states of the controlled manipulator.

The dynamics of the manipulator is represented as:

$$\tau(t) = H(q_p) \cdot \ddot{q}_p + c(q_p, \dot{q}_p) \qquad (1)$$

where:

q_p, $\dot{q}_p$ and $\ddot{q}_p$ are the generalized position, velocity and acceleration vectors respectively;

$H(q_p)$ is the generalized inertia matrix;

$c(q_p, \dot{q}_p)$ is the vector representing non-linearities due to the dynamic coupling between the links;

$\tau(t)$ is the joint-torque input vector.

Additional details of the MRAC method applied to the manipulator can be found in Sasiadek and Srinivasan.(2, 3).

4.   CONTROL OF FLEXIBLE MANIPULATORS

The MRAC scheme described earlier is well applicable to rigid manipulators. But in the case of flexible manipulators, it is not adequate to control the joint angles, in order to position the end-effector accurately. The control system must, in addition, determine the amount of deviation of the end-effector due to the deflection of the links, and correct the joint positioning accordingly.

The dynamic equations of a rigid manipulator can be used in the determination of link deflection. The various forces and moments acting on the different parts of the manipulator can be represented by an effective moment, given by the dynamic equations. Each link is then considered to be a cantilever beam fixed at its joint-end and free at its other end. A simplifying albeit conservative assumption made here is that the weight of each link (and payload, if any) is assumed to be acting through the end of that link. From classical beam theory for cantilever beams acted upon by moment loads at the end, we have the following equations for deflection and slope at each link-tip.

$$\delta_{\text{link-tip}} = -Tl^2 / 2EI_n \qquad (2)$$

$$slope_{link-tip} = -Tl/EI_n \qquad (3)$$

where:
      T is the moment load acting on the link (N-m);
      l is the link length (m);
      E is modulus of elasticity (N/m^2);
      I_n is the moment of inertia of the beam cross-section about a neutral axis, which passes through the centroid of the section.

The slope and deflection at the tip of each link are then used in the process of deflection compensation, which consists of three stages:

1. Correcting the angular position of link-1 to restore joint-2 to its specified location (Fig. 3).
2. Correcting the angular position of link-2 to overcome the deviation caused by the change in the slope at the end of link-1 (i.e. at joint-2); as shown in Fig. 4.
3. Correcting the angular position of link-2 to restore the end-effector to its target location (Fig. 5).

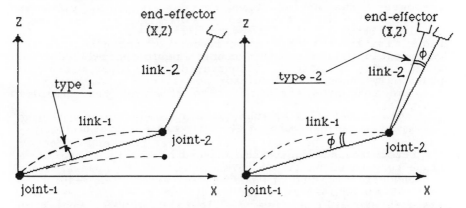

Fig. 3 Deflection Compensation   Fig. 4 Deflection Compensation

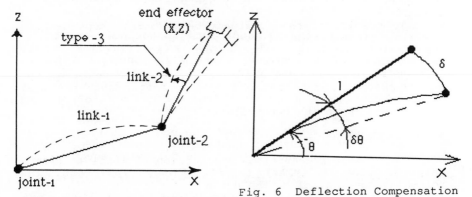

Fig. 5 Deflection Compensation,

Fig. 6 Deflection Compensation
Types 1 and 3

For steps 1 and 3 (Fig. 6), the control system updates its value of the desired position vector r_p by $(+\delta\theta)$ to $(r_p + \delta\theta)$. For step 2, the control system updates θ_2, the angle at joint-2 to $(\theta_2 + \phi)$, to compensate for the flexural slope at the end of link-1.

5.   SIMULATION RESULTS

A step input is used to study the response of the various manipulators. Such an input can be specified as follows:

$$r_p = [0.5, 1.0]$$

implying that the final joint coordinates are 0.5 and 1.0 radian. The response of the rigid manipulator to this input is shown in Fig. 7. For the purposes of comparison, a similar response of a rigid manipulator with non-adaptive control is shown in Fig. 8. While Fig. 7 shows smooth responses, the non-adaptive control (Fig. 8) produces responses with discontinuities or kinks, due to the dynamic coupling between the links. The absence of such kinks in Fig. 7 testifies to the effectiveness of the non-linearity compensation and decoupling control performed by MRAC. The insensitvity of MRAC is illustrated in Fig. 9, where the previously unloaded manipulator now carries a payload of 3 Kg. The responses are similar to those of the unloaded manipulator.

The performance of the flexible manipulator with MRAC was also investigated. A step response of the joint angles can no longer be used. Therefore, the end-effector trajectory was now monitored, instead of the joint angles. A trajectory based on simple harmonic motion was specified for the end-effector, and its positioning error with respect to that of an equivalent rigid manipulator was plotted. Two features are evident from the figure (Fig. 10). A peak of about 50 mm difference occurs at about two seconds. Secondly, an oscillating steady-state error occurs beyond 9 seconds.

To examine if this unsteady behaviour is dependant on the mass distribution of the links, another simulation of the flexible manipulator was performed with link masses of 4 and 1 Kg. (Fig. 11). This plot shows a slight reduction in maximum error, and more importantly, a convergence of the steady-state error to zero. This illustrates that the mass distribution of the links of a flexible manipulator play an important role in its controllablity with MRAC.

6.   CONCLUSIONS

This paper presents the application of model reference adaptive control (MRAC) to flexible manipulators. The following conclusions can be drawn:
1.   It is possible to control flexible manipulators effectively using MRAC;

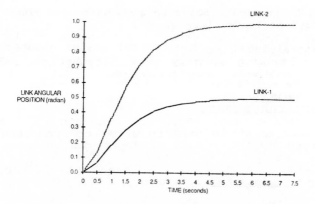

Fig. 7   Step Response of Rigid Manipulator with Adaptation

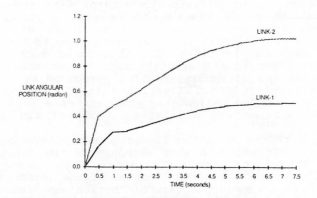

Fig. 8   Step Response of Rigid Manipulator
without Adaptation

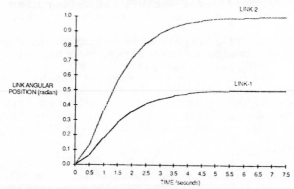

Fig. 9   Rigid Manipulator Response with Adaptation
(payload = 3 Kg.)

2. MRAC has capabilities to compensate for non-linear and coupling terms.
3. The control system based on MRAC is payload insensitive;
4. The tracking accuracy is much higher with MRAC, than with a standard control system.

7.   REFERENCES

1. Landau, I. D., 1979, 'Adaptive Control - The Model Reference Approach', Marcel Dekker Inc., New York and Basel.

2. Sasiadek, J. Z., and Srinivasan, R., 1986, <u>Proc. IFAC Int'l Symp. Theory of Robots</u>, 245 - 250.

3. Sasiadek, J. Z., and Srinivasan, R., 1987, <u>Proc. Third Canadian Univs. Conf. CAD/CAM</u>, 150 - 160.

4. Horowitz, R., and Tomizuka, M., 1980, <u>ASME J. Dyn. Syst. Meas. and Contol, 80-WA/DSC-6</u>.

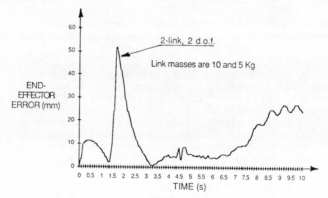

Fig. 10   End-effector Error Profile for Flexible Manipulator

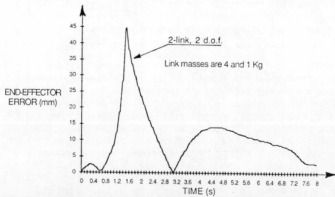

Fig. 11   End-effector Error Profile for Flexible Manipulator

Chapter 6

An approach to variable structure control of industrial robots

Z. Taha

1.0 INTRODUCTION

The design of a robot control system is made complicated by the nonlinear and coupled characteristic of it's dynamics. The dynamics of a robot can be described by a set of coupled nonlinear second order differential equations in the form of gravitational torques, Coriolis and centrifugal forces. The significance of these forces are dependent on the physical parameters of the robot, the load it carries and the speed at which the robot operates. Where accuracy is required, compensation for these parameter variations and disturbances becomes much more critical hence the design of the control system becomes much more complex.

Control strategies can generally be divided into two catogeries. Schemes that are dependent on the robot dynamic equations requires real time computation of the appropriate input generalized forces. An efficient algorithm and detailed dynamic model is hence required. Such schemes are therefore sensitive to unidentified parameter variations and disturbances. On the other hand, some robots utilizes conventional feedback scheme. Although the concept is simple, it requires difficult stability analysis. But schemes with high gain feedback or adaptive gain shows good insensitivity properties.

The theory of variable structure system (VSS) has been developed in the USSR in the last 23 years. It has found applications in the control of many processes. Essentially it is a system with discontinous feedback control. Operating such a system, in what is known as the sliding mode, makes it insensitive to parameter variations and disturbances. However two difficulties in VSS design is the derivation of the control signal from the system model. The full inverse dymamics approach [1] is extremely calculation intensive. Since variable structure systems is insensitive to parameter variations, precise system models should not be required. However this give rise to the second difficulty. When a much simplified model is used, the unidentified parameter variations and disturbances cause nonideal characteristics in the sliding mode, resulting in chattering. Utkin[2] comments that if these variations and disturbances can be measured as part of a rejection scheme, the function of the variable structure controller is reduced to the rejection of measurement error. In practice, this scheme would be expensive and difficult to implement.

In this paper an algorithm is presented which approaches these two problems. It simplifies the approach to VSS design and also results in smoother control.

2.0 BASICS OF VARIABLE STRUCTURE CONTROL.

Consider a second order system described by the canonical equations:

$$\dot{x}_1 = x_2$$
$$\dot{x}_2 = - a_2 x_2 - a_1 x_1 + bu \tag{1}$$

for which a sliding mode control is to be designed. The switching function s is chosen as

$$s = C x_1 + x_2 \tag{2}$$

that is when $s = 0$ defines the switching slope. The general reaching condition is

$$s \dot{s} < 0 \tag{3}$$

If the control law is given by the equation

$$u = \psi x_1 \tag{4}$$

where ψ is the feedback gain, the reaching condition is satisfied as follows:

$$\psi = \begin{array}{l} \beta \quad \text{if } x_1 s > 0 \\ \\ \alpha \quad \text{if } x_1 s < 0 \end{array} \tag{5}$$

Thus the design problem to be considered is the determination of the values of α, β ,and C. This will ensure that the system enters into what is called the sliding mode with the system ideally moving along the switching line. In this mode any nonlinear interactions and disturbances are rejected by the switched control signal.

3.0 ROBOT ARM MODEL

The dynamics of a robot arm can be represented by

$$T = A(Q) \; Q'' + B(Q,Q') \; Q' + C(Q) \tag{6}$$

where $A(Q)$ = coupled inertia matrix

$B(Q,Q')$ = matrix of Coriolis and centrifugal forces

$C(Q)$ = gravity matrix

T = input torques applied at various joints

For a two degree of freedom robot, under the assumption of lumped equivalent masses and massless links, the dynamics are represented by ,

$$
\begin{bmatrix} T_1 \\ T_2 \end{bmatrix} =
\begin{bmatrix} D_{11} & D_{12} \\ D_{21} & D_{22} \end{bmatrix}
\begin{bmatrix} q_1'' \\ q_2'' \end{bmatrix} +
\begin{bmatrix} D_{122}(q_2')^2 + D_{112} \, q_1' \, q_2' \\ D_{211}(q_1')^2 \end{bmatrix} +
\begin{bmatrix} D_1 \\ D_2 \end{bmatrix}
\tag{7}
$$

where

$$D_{11} = (M_1 + M_2) \, d_1^{\,2} + M_2 d_2^{\,2} + 2M_2 d_1 d_2 \cos q_2$$

$$D_{12} = M_2 d_2^{\,2} + M_2 d_1 d_2 \cos q_2$$

$$D_{21} = D_{12}$$

$$D_{22} = M_2 d_2^{\,2}$$

$$D_{112} = -2M_2 d_1 d_2 \sin q_2$$

$$D_{122} = -M_2 d_1 d_2 \sin q_2$$

$$D_{211} = D_{122}$$

$$D_1 = [(M_1 + M_2) \, d_1 \sin q_1 + M_2 d_2 \sin (q_1 + q_2)]g$$

$$D_2 = [M_2 d_2 \sin (q_1 + q_2)]g$$

For the set point regulation problem we define the state vectors

$$x^T = (e_1, x_2, e_3, x_4) = (q_1 - q_{1d}, q_1', q_2 - q_{2d}, q_2')$$

Thus equation (7) in state space representation becomes ,

$$\dot{e}_1 = x_2 \tag{8}$$

$$\dot{x}_2 = \frac{D_{22}}{d} (D_{122} \, x_2^2 + D_{112} \, x_2 \, x_4 + D_1 + T_1)$$

$$- \frac{D_{12}}{d} (D_{211} \, x_4^2 + D_2 + T_2) \tag{9}$$

$$\dot{e}_3 = x_4$$

$$x_4 = - \frac{D_{12}}{d} (D_{122} \, x_2^2 + D_{112} \, x_2 \, x_4 + D_1 + T_1)$$

$$+ \frac{D_{11}}{d} (D_{211} \, x_4^2 + D_2 + T_2) \tag{11}$$

It can be seen that synthesis of a control law of the form

$$T_i = \psi_i' \, x_1 + \psi_i^2 \, x_2 + \psi_i^3 \, e_3 + \psi_i^4 \, x_4 \tag{12}$$

would be difficult due to the nonlinear and interactive nature of the canonical equations. The objective of the approach taken in this paper is to reduce these equations to more manageable linear equations.

4.0 METHOD OF REDUCTION.

Although the physical and mathematical structure of the complete dynamic robot model are analytically coupled and non-linear, observed transient responses of robot dynamics appear to resemble transient responses of linear systems. Consequently each joint of the robot can be characterised as a single-input single-output system (SISO). The input is the actuator torque or force and the output is the joint position.

Thus the approach taken in this paper is to treat the mathematical model of the robot as a 'Black Box'. The input into this 'Black Box' is the transient response of a linear model to a step input (figure 1). The output are the motive forces or torques required by the robot to reproduce responses similar to the linear model. Samples of the input and output of the 'Black Box' are fed into an identification program which will match a low order decoupled linear time invariant model of the form

$$G(s) = \frac{Y(s)}{U(s)} = \frac{B_m s^m + B_{m-1} s^{m-1} \dots\dots B_1 s + B_o}{s^n + A_{n-1} s^{n-1} + \dots A_1 s + A_o} \qquad (13)$$

The model order m and n are selected to give the lowest possible order that will characterize the structure of the mathematical model of the robot. This can be validated by comparing the response of the model based on the identified parameters $A_0, A_1 \dots A_n$, $B_0 \dots\dots B_m$ with the desired response from the linear time invariant model, the input into the 'Black Box'.

In this paper, it was found that it was possible to reduce the dynamic model of the two link manipulator to a set of linear second order models. The values of the robot parameters used were M1=2kg, M2=5kg, d1=d2=1m. The input into the 'Black Box' is the response of

$$\ddot{y} + 2kn\dot{y} + n^2 y = n^2 y_s \qquad (14)$$

where k=1, critical damping.

to a step input. For n=5, k =1, and the initial conditions and set points were

$$q_1(0) = 0, \ q_2(0) = 0, \ q_1'(0) = q_2'(0) = 0, \ q_{1d} = 1.0, \ q_{2d} = 1.0$$

the linear model parameters of joint 1 were found to be

$$A_0 = 0.1730, \ A_1 = -0.2140, \ B_0 = 0.0265$$

and joint 2 were,

$$A_0 = 0.0438, \ A_1 = 0.3610, \ B_0 = 0.0967$$

5.0 CONTROLLER DESIGN

The reduced order model of the manipulator can be represented in canonical form as

$$\dot{e}_1 = x_2$$

$$\dot{x}_2 = B_o^1 T_1 - A_1^1 x_2 - A_o^1 x_1$$

$$\dot{e}_3 = x_4$$

$$\dot{x}_4 = B_o^2 T_2 - A_1^2 x_4 - A_o^2 x_3$$

The switching lines are given by,

$$s_1 = C\,e_1 + x_2$$

$$s_2 = C\,e_3 + x_4$$

The control law $T_i = \psi\,e_i$ is chosen. For joint 1,

$$\dot{s}_1 = C\,\dot{e}_1 + \dot{x}_2 = C\,x_2 + B_o^1 T - A_1^1 x_2 - A_o^1 x_1$$

But in the proximity of the switching line $x_2 = -Ce_1$. Therefore

$$\dot{s}_1 = -\,C^2 e_1 + B_o^1 \psi e_1 + A_1^1 C e_1 - A_o^1 x_1$$

The general reaching condition $s_1 \dot{s}_1 \le 0$, that is

$$[-C^2 + B_o^1 \psi + A_1^1 C]s_1 e_1 - A_o^1 x_1 s_1 < 0$$

Therefore

$$\psi = \begin{array}{ll} < \varepsilon & \text{if } s_1 e_1 > 0 \\ > \varepsilon & \text{if } s_1 e_1 < 0 \end{array}$$

where

$$\varepsilon = \frac{C^2 - A_1^1 C}{B_o^1}$$

Therefore, the general reaching condition may be satisfied by

$$\psi = \begin{array}{ll} \beta & \text{if } s_1 e_1 > 0 \\ \alpha & \text{if } s_1 e_1 < 0 \end{array}$$

where β is negative and α is positive and their magnitudes are greater then ε .

6.0 SIMULATION AND CONCLUSIONS

A simulation of the two link manipulator have shown that the transient responses produced by the controller are smooth and effectively decoupled (figure 2). The behaviour of the control signals clearly indicates that a sliding motion occurs as intended. The control signals are smoother with evidently less chattering (figure 3).

It has been shown that the complex and nonlinear dynamics of a robot can be reduced to a set of linear models. This has made the design of variable structure controller less complex. It has also improved the controller resulting in smoother control signals and transient responses.

REFERENCES

1. K.K.D.Young, 'Controller Design for a Manipulator Using Theory of Variable Structure Systems', IEEE Transactions on Systems, Man, and Cybernetics, Vol.SMC-8, No.2, Febuary 1978.

2. N.N.Bengiamin, B.Kauffmann, 'Variable Structure Position Control', Control Systems Magazine, August, 1984.

3. P.J.Drazan, Z.Taha, 'Decoupled control of Robots', 16th International Symposium on Industrial Robots, 30 September-2 October 1986, Brussels, Belgium.

4. Z.Taha, Dynamics and Control of Robots, PhD Thesis, University of Wales, 1987.

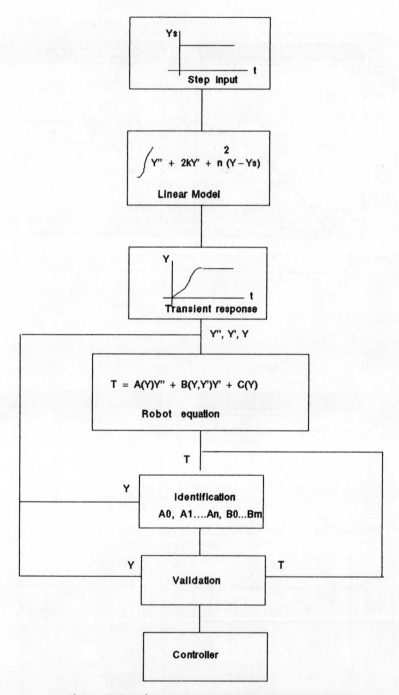

Fig.1 Reduction process of robot dynamic model

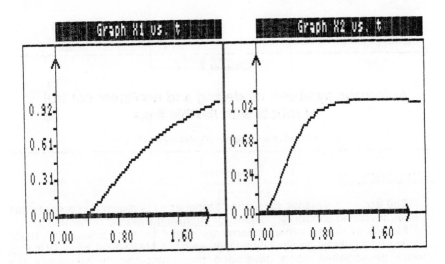

Fig.2 Transient responses of joint 1 (x_1) and joint 2 (x_2)

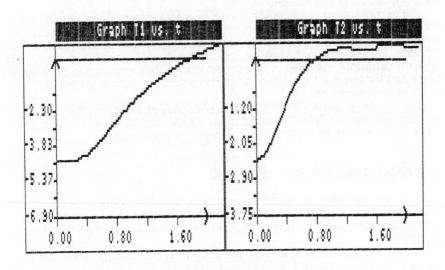

Fig.3 Control signals T_1 and T_2 of joint 1 and joint 2
respectively

Chapter 7

Automatic symbolic modelling and nonlinear control of robots with flexible links

A. De Luca, P. Lucibello and F. Nicolò

1. INTRODUCTION

The accurate dynamic modeling of robot arms is required in order to use nonlinear advanced control techniques for improving dynamic performances.

Models in closed symbolic form are useful for control analysis. Using symbolic and algebraic manipulation languages like MACSYMA or REDUCE, a software packagehas been developed that automatically generates the dynamic model for robots with rigid and/or flexible links, including joint elasticity (Cesareo et al. (1), Lucibello et al. (2)).

As far as link flexibility is concerned, the kinematic behaviour of a flexible arm can be modeled using a finite number of functions which describe the link deformation. The Lagrangian approach is followed to derive the dynamic equations. Other approaches to modeling can be found in Nicosia et al. (3), Book (4), and Judd and Falkenburg (5).

For rigid arms and elastic joints, several control algorithms using nonlinear static or dynamic state-feedback have been proposed (Bejczy (6), Freund (7), Tarn et al. (8), De Luca et al. (9), Marino and Spong (10)). These methods are essentially based on the input-output decoupling of the robot dynamic behaviour. As a result, the closed-loop system turns out to be a set of linear and independent subsystems, so that standard techniques can be used for robot motion control.

In this paper we investigate the possibility of extending this approach to robots with flexible links. We consider as a case study a two-link planar robot with rotational joints and with only the second link being flexible.

2. KINEMATIC AND DYNAMIC MODELING

We consider open mechanical chains of rigid and/or flexible links interconnected by lower kinematic pairs. No specific restrictions on the deformability of the elastic links is imposed, e.g. the link is not required to be a planar beam.

For a generic link, let r be the position vector of a generic point P of the link in its undeformed configuration, relative to an ortonormal frame R attached to the joint preceding the link. The position vector x of P relative to an absolute frame R_0 attached to the base can be written as

$$x = x_0 + Q(r + u)$$

where x_0 is the position vector of the origin of R relative to R_0, Q is the matrix which rotates R_0 over R, and u is the elastic displacement of P. u is a function of r and time t.

In order to work with a finite dimensional model, let us assume that the vector function u can be described by

$$u(r,t) = F(r) \ p(t)$$

where F is a 3xH matrix of allowed deformation modes of the link and p is a H-vector of modes amplitudes; H is the number of deformation modes.

The computation of x_0 and Q is performed recursively from one joint to the next one; the generic step can be splitted in: 1) a joint rotation (or translation) β; 2) a roto-translation due to the deformation of link, function of the vector p only.

Let the joint variables and the elastic variables be aggregated into a vector q. Taking the time derivative of the kinematic relation

$$x(q) = x_0(q) + Q(q) \ (\ r + u(r,t) \ )$$

yields the velocity of the generic point:

$$\dot{x}(q,\dot{q}) = \dot{x}_0 (q,\dot{q}) + \dot{Q}(q,\dot{q}) \ (\ r + u(r,t) \ ) + Q(q) \ \dot{u}(r,t)$$

The total kinetic energy of the system is the sum of the kinetic energies T of each link

$$T = \frac{1}{2} \int_L \dot{x}^T \dot{x} \ dm$$

where L denotes link mass. Substituting the velocity expression in T gives

$$T = \frac{1}{2} \int_L [\ \dot{x}_0^T \dot{x}_0 + 2 \ \dot{x}_0^T \ \dot{Q} \ (r + u) + 2 \ \dot{x}_0^T \ Q \ \dot{u}$$

$$+ (r + u)^T \ \dot{Q}^T \ \dot{Q} \ (r + u) + 2 \ \dot{u} \ Q^T \dot{Q} \ (r + u) + \dot{u}^T Q^T Q \ \dot{u} \ ] \ dm$$

In (2), it is shown that only the following inertial data are needed to write down the kinetic energy of a link:

$$m = \int_L dm \qquad \text{total link mass}$$

$$SR = \int_L r \ dm \qquad \text{static mass moment}$$

$$JR = \int_L r\, r^T\, dm \qquad \text{Euler matrix}$$

$$SE = \int_L F\, dm \qquad \text{elasto-static mass moment matrix}$$

$$M = \int_L F^T F\, dm \qquad \text{mass matrix}$$

$$JM_{ijh} = \int_L r_i\, F_{jh}\, dm \qquad \text{elements of the 3rd order mixed inertial tensor}$$

$$JE_{ijhk} = \int_L F_{ij}\, F_{hk}\, dm \qquad \text{elements of the 4th order elastic inertial tensor}$$

The code uses this list of symbols as input data. With the chosen formulation, the above quantities can be defined both for concentrated and distributed mass systems. Of course, it is irrelevant how the numerical values for these quantities are obtained. According to the specific link structure, the appropriate computing procedure may be selected (e.g. using finite elements).

The total gravitational energy of the system is the sum of the gravitational energies U_g of each link

$$U_g = -\int_L g^T x\, dm = -m\, g^T x_0 - g^T Q\, SR - g^T Q\, SE\, p$$

We assume that the deformable links undergo elastic deformations only. The link elastic energy $U_e(p)$ can be computed by any suitable program for structural analysis. In particular, for the elastic linear case

$$U_e(p) = \frac{1}{2}\, p^T K\, p$$

and the so-called stiffness matrix K has to be determined.

Finally, the code incorporates also a Rayleigh-type dissipative function

$$R = \frac{1}{2}\, \dot{p}^T C\, \dot{p}$$

where C is the so-called damping matrix. This allows to simulate a viscous behaviour of the deformable link.

Using Lagrange equations, we obtain the dynamic model in the form:

$$B(q)\, \ddot{q} + n(q,\dot{q}) = u(t)$$

where $B(q)$ is the generalized matrix of inertia and $n(q,q')$ are the Coriolis, gravitational, elastic and dissipative forces. For an N jointed arm, vector $u(t)$ has

N entries given by the torques (or forces) applied to the joints and H entries equal to zero.

The present version of the modeling software can handle up to six flexible links interconnected by rotational or prismatic joints, provided enough computing power is available. Elasticity concentrated at the joints may also be included. In addition, an interface with standard numerical integration routines is operative on a Univac 1100 mainframe.

3. NONLINEAR CONTROL TECHNIQUES

For rigid robot arms (8) and arms with elasticity concentrated at the joints (9), nonlinear control techniques have been successfully used. These have been sometimes labeled in the robotic field as inverse dynamics or computed torque methods (6) and have been used in both joint and cartesian space control design.

For a given nonlinear system of the form

$$\mathbf{x} = f(\mathbf{x}) + g(\mathbf{x})\,\mathbf{u} \qquad \mathbf{y} = h(\mathbf{x})$$

$\mathbf{x} \in R^n$, $\mathbf{u} \in R^m$, $\mathbf{y} \in R^m$, these approaches are generally based on the use of state feedback to achieve input-output decoupling and exact state linearization in the closed-loop. Doing so, the control problem is reduced to a set of linear single-input single-output ones.

In particular, a static feedback of the form

$$\mathbf{u} = \alpha(\mathbf{x}) + \beta(\mathbf{x})\,\mathbf{v}$$

with full control (i.e. $\beta(\mathbf{x})$ nonsingular), achieves noninteraction if and only if the so-called decoupling matrix $A(\mathbf{x})$ is nonsingular (Isidori et al.(11)), where the elements $a_{ij}(\mathbf{x})$ of this matrix are

$$a_{ij}(\mathbf{x}) = L_{g_j} L_f^{(r_i - 1)} h_i(\mathbf{x})$$

and r_i is the relative degree of the i-th output. $L_f^k h$ is the k-th Lie derivative of the function h w.r.t. the vector field f. In this case, a possible decoupling law uses

$$\beta(\mathbf{x}) = A^{-1}(\mathbf{x}), \qquad \alpha(\mathbf{x}) = \beta(\mathbf{x}) \cdot \text{col}\{ L_f^{r_i} h_i(\mathbf{x}) \}$$

The resulting system has an input-output behaviour which is linear in the proper local coordinates. An additional unobservable part, a sink, may exist in the closed-loop. In general, the stability of this possibly nonlinear part has to be investigated in order to conclude that the decoupling law is a satisfactory one. However, if

$$\sum_{i=1}^{m} r_i = n$$

this sink does not exist and the decoupling feedback is also a fully linearizing one. This is the case of rigid robot arms.

If the decoupling matrix is structurally singular, the same results may still be obtained considering the more general class of nonlinear dynamic state feedback of the form

$$\dot{z} = a\,(x,z) + b(x,z)\,v\,, \qquad u = c(x,z) + d(x,z)\,v$$

where z is the state of the dynamic compensator. This is what is required for instance for full linearization (and decoupling) of robots with elasticity concentrated at the joints (De Luca (12)).

In the following we will investigate the applicability of these techniques for the control of a flexible arm.

4. A CASE STUDY

We consider a two-link rotational robot, moving on a horizontal plane, whose second link is flexible. The dynamic model is reported in the Appendix and includes only one flexible mode (Fig.1).

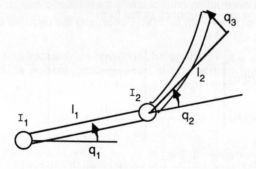

Fig.1 A two-link planar arm with a flexible second link

The control aim is to let the end effector track a given reference trajectory. Let $x = (q,q') \in R^6$. The output of the system is defined as

$$\mathbf{y} = \begin{bmatrix} q_1 \\ q_2 + \dfrac{q_3}{l_2} \end{bmatrix} = \mathbf{h}(\mathbf{x})$$

Note that the second component of the output is the linearized expression of the angular position of the tip with respect to the first link axis.

The associated decoupling matrix is then

$$A(\mathbf{x}) = L_g L_f h(\mathbf{x}) = \begin{bmatrix} d_{11} & d_{21} \\ d_{21} + \dfrac{d_{31}}{l_2} & d_{22} + \dfrac{d_{31}}{l_2} \end{bmatrix}$$

where d_{ij} are elements of $B^{-1}(\mathbf{q})$, the inverse of the generalized inertia matrix. The relative degrees of the two outputs are $r_1 = r_2 = 2$. The determinant of $A(\mathbf{x})$ is

$$\det A(\mathbf{x}) = \frac{1}{\det B(\mathbf{q})} [\, b_{33} - \frac{b_{23}}{l_2} \,] \neq 0$$

where the last inequality follows from the expressions reported in Appendix.

The chosen decoupling control law is then a static one, $\mathbf{u} = \alpha(\mathbf{x}) + \beta(\mathbf{x})\mathbf{v}$, with $\beta(\mathbf{x}) = A^{-1}(\mathbf{q})$, $\alpha(\mathbf{x}) = A^{-1}(\mathbf{q}) L_f^2 h(\mathbf{x})$. Performing the required calculations

$$\mathbf{u} = \begin{bmatrix} n_1 (\mathbf{q},\dot{\mathbf{q}}) \\ n_2 (\mathbf{q},\dot{\mathbf{q}}) \end{bmatrix} + \frac{n_3 (\mathbf{q},\dot{\mathbf{q}})}{b_{33} - \dfrac{b_{23}}{l_2}} \begin{bmatrix} \dfrac{b_{12}(\mathbf{q})}{l_2} - b_{13}(\mathbf{q}) \\ \dfrac{b_{22}(\mathbf{q})}{l_2} - b_{23}(\mathbf{q}) \end{bmatrix} + A^{-1}(\mathbf{q}) \mathbf{v}$$

A sink of order $n-(r_1+r_2) = 2$ exists. The closed-loop dynamics can be written as

$$\ddot{\mathbf{y}} = \mathbf{v}$$

for the input-output decoupled part, and

$$\ddot{q}_3 = \frac{20}{m_2} n_3 (\mathbf{y}, \dot{\mathbf{y}}, q_3, \dot{q}_3)$$

$$+ [\, \frac{20}{3} l_1 \cos (y_2 - \frac{q_3}{l_2}) + 5 l_2 \,] v_1 + 5 l_2 v_2$$

for the sink, where the following set of coordinates has been used

$$\{y_1, \dot{y}_1, y_2, \dot{y}_2, q_3, \dot{q}_3\}.$$

The behaviour of the closed-loop system is unstable for $y = 0$, $y' = 0$, $v = 0$. In this case, the dynamics of the sink becomes

$$\ddot{q}_3 = \frac{20}{m_2} n_3 (0, 0, q_3, \dot{q}_3) = [\frac{20 K}{m_2} - \frac{4}{l_2^2} \dot{q}_3^2] q_3$$

which is the so-called zero dynamics associated to the system (Isidori and Moog (13)), i.e. the dynamics left in the system when the input is chosen in such a way to force a constant zero output.

We may try to design the new input v in such as to remove this instability. However, if noninteraction and/or linear input-output behaviour has to be preserved, such feedback design for v does not exist. It can also be shown that this unstable mode is a *fixed* one (Isidori and Grizzle(14)) with respect to all possible decoupling laws u.

5. CONCLUSIONS

The automatic symbolic procedure for deriving dynamic models of flexible robot arms has been proven to be useful in order to investigate some control properties of this class of systems.

In particular, we have shown that decoupling with internal stability is not achievable for a two-link planar arm with one flexible link, when the end effector position is the objective of control.

This nonlinear analysis in the time domain seems to parallel the experimental findings of Canudas and Van den Bossche (15) and of Cannon and Schmitz (16) on the linearized model of a one-link flexible arm. In fact, they found a non-minimum phase zero in the input-output frequency response that, closing the feedback loop, may lead to an unstable pole.

Exploiting further the above parallel, one may think to overcome this restrictions by letting the input-output behaviour of the original nonlinear system be matched via feedback (see e.g. De Luca et al.(17)) to a linear transfer matrix with a non-minimum phase zero.

REFERENCES

1. Cesareo, G., Nicolò, F., Nicosia, S., 1984, DYMIR : a code for generating dynamic model of robots, IEEE Int. Conf. on Robotics and Automation, Atlanta.
2. Lucibello, P., Nicolò, F., Pimpinelli, R., 1986, Automatic symbolic modelling of robots with a deformable link, IFAC Int. Symp. on Theory of Robots, Vienna.
3. Nicosia, S., Tomei, P., Tornambè, A., 1986, Dynamic modelling of flexible manipulators, IEEE Int. Conf. on Robotics and Automation, S.Francisco.

4. Book, W.J., 1984, Recursive Lagrangian dynamics of flexible manipulator arms, Int. J. of Robotics Research, 3, 87-101.
5. Judd, R.P., Falkenburg, D.R., 1985, Dynamics of nonrigid articulated robot linkages, IEEE Trans. Autom. Contr., 30, 499-502.
6. Bejczy, A.K., 1974, Robot arm dynamics and control, Jet Propulsion Lab, California Institute of Technology, TM 33-669.
7. Freund, E.K., 1982, Fast nonlinear control with arbitrary pole-placement for industrial robots and manipulators, Int. J. of Robotics Research, 1, 65-78.
8. Tarn, T.J., Bejczy, A.K., Isidori, A., Chen, Y., 1984, Nonlinear feedback in robot arm control, IEEE Conf. on Decision and Control, Las Vegas.
9. De Luca, A., Isidori, A., Nicolò, F., 1985, Control of robot arms with elastic joints via nonlinear dynamic feedback, IEEE Conf. on Decision and Control, Ft.Lauderdale.
10. Marino, R., Spong, M.W., 1986, Nonlinear control techniques for flexible joint manipulators: a single link case study, IEEE Int. Conf. on Robotics and Automation, S.Francisco.
11. Isidori, A., Krener, A.J., Gori Giorgi, C., Monaco, S., 1981, Nonlinear decoupling via feedback: a differential geometric approach, IEEE Trans. Autom. Contr., 26, 331-345.
12. De Luca, A., 1987, Control properties of robot arms with joint elasticity, Int. Symp. on Math. Theory of Networks and Systems, Phoenix.
13. Isidori, A., Moog, C.H., 1987, On the nonlinear equivalent of the notion of transmission zeros, 'Modeling and Adaptive Control', C.I.Byrnes,A.H.Kurszanski Eds., Springer Verlag.
14. Isidori, A., Grizzle, J.W., 1987, Fixed modes and nonlinear interacting control with stability, submitted for publication.
15. Canudas de Wit, C., Van den Bossche, E., 1986, Adaptive control of a flexible arm with explicit estimation of the payload mass and friction, IFAC Int. Symp. on Theory of Robots , Vienna.
16. Cannon, R.H., Schmitz, E., 1984, Initial experiments on the end-point control of a flexible one-link robot, Int. J. Robotics Research, 3, 62-75.
17. De Luca, A., Isidori, A., Nicolò, F., 1985, An application of nonlinear model matching to the control of robot arm with elastic joints, IFAC Symp. on Robot Control, Barcelona.

<u>APPENDIX</u>

The dynamic terms of the considered two-link planar elastic robot are reported here. With reference to Fig.1, q_1 and q_2 are the joint rigid rotations while $q_3 = p$ is the parameter of the single assumed parabolic mode of vibration of the second link.

For the symmetrical inertia matrix $B(\mathbf{q})$, we have:

$$b_{11}(\mathbf{q}) = b_{111} q_3 \sin q_2 + b_{112} \cos q_2 + b_{113} q_3^2 + b_{114}$$

$$b_{12}(\mathbf{q}) = b_{121} q_3 \sin q_2 + b_{122} \cos q_2 + b_{123} q_3^2 + b_{124}$$

$$b_{13}(\mathbf{q}) = b_{131} \cos q_2 + b_{132}$$

$$b_{22}(\mathbf{q}) = b_{221} q_3^2 + b_{222}$$

$$b_{23} = \frac{1}{4} m_2 l_2$$

$$b_{33} = \frac{1}{5} m_2$$

where the following coefficients have been introduced:

$$b_{111} = -\frac{2}{3} m_2 l_1 \qquad b_{112} = m_2 l_1 l_2 \qquad b_{113} = \frac{1}{5} m_2$$

$$b_{114} = I_1 + I_2 + m_2 (l_1^2 + \frac{l_2^2}{3})$$

$$b_{121} = -\frac{1}{3} m_2 l_1 \qquad b_{122} = \frac{1}{2} m_2 l_1 l_2 \qquad b_{123} = \frac{1}{5} m_2$$

$$b_{124} = I_2 + \frac{1}{3} m_2 l_2^2 \qquad b_{131} = \frac{1}{3} m_2 l_1 \qquad b_{132} = \frac{1}{4} m_2 l_2$$

$$b_{221} = \frac{1}{5} m_2 \qquad b_{222} = I_2 + \frac{1}{3} m_2 l_2^2$$

The dynamic terms in $\mathbf{n}(\mathbf{q},\mathbf{q}')$ can be computed from the elements $b_{ij}(\mathbf{q})$ of the inertia matrix and the knowledge of potential and dissipative energies. No structural damping is included in the considered model. Using the Einstein summation convention, we have:

$$n_i(\mathbf{q},\dot{\mathbf{q}}) = b_{ij,k}(\mathbf{q}) \dot{q}_j \dot{q}_k - \frac{1}{2} b_{hk,i}(\mathbf{q}) \dot{q}_h \dot{q}_k + U_{g,i}(\mathbf{q}) + U_{e,i}(\mathbf{q})$$

where the comma denotes derivative w.r.t. the homologous component of $\mathbf{q}$. In this case, $U_g = 0$ while the only contribution of the elastic energy is a term Kq_3 in $n_3(\mathbf{q},\mathbf{q}')$.

Chapter 8

On the complexity reduction of the coriolis and centripetal effects of a 6-DoF robot manipulator

A. Y. Zomaya and A. S. Morris

ABSTRACT:-

The equations used in calculating the different forces and
torques which control the movement of a robot manipulator
involve a considerable amount of differential and non-linear
terms that possess high computational complexity. Centripetal and Coriolis effects are of great importance when the
manipulator is moving at high speeds. The previous effects,
based on the Lagrangian formulation, have been simplified
and a lower order form produced which has reduced computational complexity. Simulation results for a robot arm have
been obtained to check for the validity of the derivation.

INTRODUCTION:

   Robot arm dynamics and mechanisms deals mainly with the
mathematical formulation of the equations of robot manipulator motion, that is, computing the actuators torques and
forces to give a lower-pair kinematic chain certain desired
trajectories (inverse dynamics), or, given the forces and
torques to calculate the accelerations and velocities of the
robot arm joints (forward dynamics). The dynamics consists
of a set of differential, non-linear, and matrix oriented
equations which describes the behaviour of the robot arm and
allows for great flexibility in computer dynamic modelling
and simulation studies to evaluate and apply the different
control and analysis schemes.

   In recent years various techniques and approaches have
been developed to formulate robot dynamics. The Lagrangian
|1,2,3,4,14| has low computational efficiency with equations
of order $O(n^4)$, but otherwise is a well organized and systematic method which give good insight into the application of
different control techniques. The generalized D'Alembert
formulation |5| has an order of $O(n^3)$ and consists of a
fairly well structured set of equations. The Newton-Euler
formulation |6,8,9,15| has a messy derivation but is the
most efficient formulation with equations of order $O(n)$ that
follow from its vector structure and recursive nature.
Tabulation-dependent techniques, such as the configuration
space method |12|, have very serious difficulties owing to
their enormous computer memory requirements.

Other approaches include dynamic equations of Kane $|13|$,
and the use of parallel processing and multi-tasking to re-
duce the order of computations $|16,17,18,19|$. Of the
previous methods, the most commonly used are the Lagrangian
and Newton-Euler. The interaction and equivalence between
these has been shown by Silver$|11|$.

The Lagrangian formulation of the dynamic equations has a
simple and algorithmic representation obtained from Lagrangian mechanics. The set of equations consists of second-
order, highly-coupled, non-linear differential expressions
which can be written in a compact form:

$$
F_i = \sum_{k=i}^{n} \sum_{j=1}^{n} \mathrm{Tr}\left\{ \frac{\partial H_o^k}{\partial v_i} J_k \left|\frac{\partial H_o^k}{\partial v_i}\right|^T \right\} \ddot{v}_j + \sum_{r=i}^{n} \sum_{j=1}^{n} \sum_{k=1}^{n} \mathrm{Tr}\left\{ \frac{\partial^2 H_o^r}{\partial v_j \partial v_k} \right.
$$

$$
J_r \left|\frac{\partial H_o^r}{\partial v_i}\right|^T \right\} \dot{v}_j \dot{v}_k - \sum_{j=1}^{n} m_j \ g\left\{\frac{\partial H_o^j}{\partial v_i}\right\} r_j \ , \quad i=1,2,\ldots,n \qquad (1)
$$

where

H_o^i: link transformation matrices

J_r: link inertia matrix

g :   gravitational effects vector

Tr: trace operator of a matrix $(\mathrm{Tr}(A) = \sum_{i=1}^{m} a_{ii})$.

F_i: force (prismatic joint) or torque (revolute joint)
     acting at joint (i).

$v_i, \dot{v}_i, v_i$,:   position, velocity and acceleration of joint
          (i)

r_j: centre of mass of link j according to its own
     coordinates.

m_j: mass of link j

n :   degree of freedom (DoF).

Eq.(1) was applied to a six DoF robot arm (stanford arm) and
the different terms were analyzed and shown to be intertia
loading, coupling coriolis and centripetal reactions, and
the gravity effects. The inertial and gravity terms are of
particular significance in controlling the servo stability
and positioning accuracy of the robot arm. The coriolis and
centripetal forces are important in high speed movements.
All the attempts mentioned previously were made to solve for
the dynamic equations with special interest in the coriolis
effects, but some of the researchers neglected these second
order effects under the assumption of low speed movements,
and this assumption led to a suboptimal dynamic performance
because of speed restrictions. In this paper, a simpli-
fication is described for the mathematical representation,
and accordingly for the computational complexity of the
coriolis and centripetal forces of a six DoF robot arm.

This is achieved by reducing the order of the lagrangian
representation of those forces. Results obtained from simulation programs for the Stanford manipulator are given.

The Lagrangian:

The use of the Lagrangian formulation has the advantage of
deriving the mechanics and dynamics of complex systems in
a well structured and organised manner but it is very difficult to utilize this in real-time control without simplifi-
cation. The symbolism and matrix notations used in $|1,3|$
which depend on the Denavit-Hartenberg representation $|10|$
will be used in our discussion. Eq.(1) that controls and
governs the motion of the robot arm might be written in an
alternate form:

$$F_i = \sum_{j=1}^{n} P_{ij} \ddot{q}_j + \sum_{j=1}^{n} \sum_{k=1}^{n} P_{ijk} \dot{q}_j \dot{q}_k + P_i \tag{2}$$

$$i = 1,2,\ldots n$$

where

$\quad P_{ii}$, effective inertia at joint (i)

$\quad P_{ij}$, coupling inertia between joint (i) and (j)

$$P_{ij} = \sum_{\ell=\max(i,j)}^{n} Tr\left(\frac{\partial H\ell}{\partial q_j} \, J_\ell \, \left|\frac{\partial H\ell}{\partial q_i}\right|^T\right) \tag{3}$$

$\quad P_{ijj}$, Centripetal forces at joint (i) due to velocity
$\qquad$ at joint (j).

$\quad P_{ijk}$, Coriolis forces at joint (i) due to velocities at
$\qquad$ joint (j) and (k).

$$P_{ijk} = \sum_{\ell=\max(i,j,k)}^{n} Tr\left(\frac{\partial^2 H\ell}{\partial q_j \, q_k} \, J_\ell \, \left|\frac{\partial H\ell}{\partial q_i}\right|^T\right) \tag{4}$$

$\quad P_i$, gravity loading vector

$$P_i = \sum_{\ell=1}^{n} -m_\ell \, g^T \left(\frac{\partial H_\ell}{\partial q_i}\right) {}^\ell r_\ell \tag{5}$$

and

$$\frac{\partial H_\ell}{\partial q_i} = H_\ell \, {}^\ell\Delta_i \tag{6}$$

where ${}^\ell\Delta_i$ is the differential translation and rotation$_{th}$
transformation matrix if joint ℓ with respect to the i
joint coordinate given by

$$
{}^{\ell}\Delta_i = \begin{vmatrix} 0 & -{}^{\ell}\delta_{iz} & {}^{\ell}\delta_{iy} & {}^{\ell}d_{ix} \\ {}^{\ell}\delta_{iz} & 0 & -{}^{\ell}\delta_{ix} & {}^{\ell}d_{iy} \\ -{}^{\ell}\delta_{iy} & {}^{\ell}\delta_{ix} & 0 & {}^{\ell}d_{iz} \\ 0 & 0 & 0 & 0 \end{vmatrix} \tag{7}
$$

and J_i is a pseudo inertia matrix, where the elements composing the matrix are the moments of inertia, cross product of inertia and first moments of each link, i.e.

$$
J_i = \begin{vmatrix} \dfrac{-I_{xxi}+I_{yyi}+I_{zzi}}{2} & I_{xyi} & I_{xzi} & m_{ixi} \\[2mm] I_{xyi} & \dfrac{I_{xxi}-I_{yyi}+I_{zzi}}{2} & I_{yzi} & m_{iyi} \\[2mm] I_{zyi} & I_{yzi} & \dfrac{I_{xxi}+I_{yyi}-I_{zzi}}{2} & m_{izi} \\[2mm] m_{ixi} & m_{iyi} & m_{izi} & m_i \end{vmatrix} \tag{8}
$$

Inertial and gravity terms:

The effective and coupling inertia terms at eq.(3) have been shown in |1| to be:

$$
P_{ij} = \sum_{\ell=\max i,j}^{n} Tr({}^{\ell}\Delta_j \ J_{\ell} \ {}^{\ell}\Delta_i^{T}) \tag{9}
$$

and the gravity loading vector are given as:

$$
P_i = {}^{i-1}g \sum_{\ell=i}^{n} m_{\ell} \ {}^{i-1}r_{\ell} \tag{10}
$$

where

$$
{}^{i-1}g = \begin{cases} \{-g.o & g.n & 0 & 0\} & \text{rotational joint} \\ \{0 & 0 & 0 & -g.a\ \} & \text{prismatic joint} \end{cases}
$$

The symmetry of the matrix P_{ij} was also shown in |1,3| which led to $p_{ij}=p_{ji}$. For more detailed discussion about the previous derivations, the reader is referred |1,3|.

Coriolis and Centripetal effects:

These terms of great importance in high speed operations which is the case in a lot of industrial applications. Eq. (4) can be simplified to give a reduced model of low computational complexity.

According to the mathematical identity.

$$\frac{\partial^2 A}{\partial x \partial y} = \frac{\partial}{\partial x} \left(\frac{\partial A}{\partial y}\right) \quad , \quad A: \text{ matrix} \ ; \ x,y \text{ scalar variables}$$

Eq.(4) could be manipulated as follows;

$$\frac{\partial H_\ell}{\partial q_i} = H_\ell{}^\ell\Delta_i \ , \quad \left(\frac{\partial H_\ell}{\partial q_i}\right)^T = {}^\ell\Delta_i^T \ H_\ell^T \tag{11}$$

So,

$$\frac{\partial^2 H_\ell}{\partial q_j \partial q_k} = \frac{\partial}{\partial q_j} \left(\frac{\partial H_\ell}{\partial q_k}\right) = \frac{\partial}{\partial q_j} \ (H_\ell{}^\ell\Delta_k)$$

which if expanded more gives ;

$$\frac{\partial^2 H_\ell}{\partial q_j \partial q_k} = \left(\frac{\partial H_\ell}{\partial q_j}\right) {}^\ell\Delta_k + H_\ell\left(\frac{\partial {}^\ell\Delta k}{\partial q_j}\right) \tag{12}$$

The second term of eq.(12) could be neglected because of its small significance in effecting the accuracy of the calculations, which yields,

$$\frac{\partial^2 H_\ell}{\partial q_j \partial q_k} = \frac{\partial H_\ell}{\partial q_j} \ {}^\ell\Delta_k \tag{13}$$

Substituting (6) into (13),

$$\frac{\partial^2 H_\ell}{\partial q_j \partial q_k} = H_\ell{}^\ell\Delta_j{}^\ell\Delta_k \tag{14}$$

now substituting (11), (14) into eq.(4) gives a better form for simulation purposes;

$$P_{ijk} = \sum_{\ell=\max i,j,k}^{n} \text{Tr} \ (H_\ell{}^\ell\Delta_j{}^\ell\Delta_k J_\ell{}^\ell\Delta_i^T H^T{}_\ell) \tag{15}$$

Eq.(15) could be simplified further; premultiplying and post multiplying by H_ℓ and $H^T{}_\ell$ respectively will effect the rotation part only, hence the trace will remain unchanged.

Eq.(15) will reduce to:

$$P_{ijk} = \sum_{\ell=\max i,j,k}^{n} ({}^\ell\Delta_j{}^\ell\Delta_k J_\ell{}^\ell\Delta_i^T) \tag{16}$$

now expanding the expression; and assuming a matrix (M) such that:

$$M = {}^\ell\Delta_j{}^\ell\Delta_K J_\ell{}^\ell\Delta_i^T$$

the matrix (M) will have the following form:

$$M = \begin{vmatrix} m_{11} & m_{12} & m_{13} & \emptyset \\ m_{21} & m_{22} & m_{23} & \emptyset \\ m_{21} & m_{32} & m_{33} & \emptyset \\ \emptyset & \emptyset & \emptyset & \emptyset \end{vmatrix}$$

the important elements of the previous matrix are the diagonal elements which constructs the trace operator. So eq. (16) will reduce to:

$$P_{ijk} = \sum_{\ell=\max i,j,k}^{n} (m_{11}+m_{22}+m_{33}) \tag{17}$$

If in matrix $|J_i|$, $I_{xy_i} = I_{xz_i} = I_{yz_i} = \emptyset$ (which is the case in a lot of robot manipulators, the elements of eq.(17) will be as follows:

$$m_{11} = Q_1 + Q_2 + Q_3$$

where,

$$Q_1 = (-{}^\ell\delta_{iz})\{{}^\ell\delta_{jy}\,{}^\ell\delta_{kx}\left|\frac{I_{xx\ell}-I_{yy\ell}+I_{zz\ell}}{2}\right| + |{}^\ell d_{ky}(-{}^\ell\delta_{jz})+{}^\ell\delta_{jy}$$
$${}^\ell d_{kz}|m_\ell y_\ell$$

$$Q_2 = ({}^\ell\delta_{iy})\{{}^\ell\delta_{jz}\,{}^\ell\delta_{kx}\left|\frac{I_{xx\ell}+I_{yy\ell}-I_{zz\ell}}{2}\right| + (-{}^\ell\delta_{jz}){}^\ell d_{ky}+{}^\ell\delta_{jy}$$
$${}^\ell d_{kz}\,m_\ell z_\ell$$

$$Q_3 = \{|(-{}^\ell\delta_{jz}){}^\ell\delta_{kz}+{}^\ell\delta_{jy}(-{}^\ell\delta_{ky})|X_\ell+{}^\ell\delta_{jy}\,{}^\ell\delta_{kx}y_\ell+{}^\ell\delta_{jz}\,{}^\ell\delta_{kx}z_\ell +$$
$$|(-{}^\ell\delta_{jz}){}^\ell d_{ky}+{}^\ell\delta_{jy}\,{}^\ell d_{kz}|\}^\ell d_{ix}m_\ell$$

$$m_{22} = Q_4 + Q_5 + Q_6$$

where,

$$Q_4 = ({}^\ell\delta_{iz})\{{}^\ell\delta_{ky}\,{}^\ell\delta_{jx}\left|\frac{-I_{xx\ell}+I_{yy\ell}+I_{zz\ell}}{2}\right| + |{}^\ell\delta_{jz}\,{}^\ell d_{kx}-{}^\ell\delta_{jx}\,{}^\ell d_{kz}|$$
$$m_\ell x_\ell\}$$

$$Q_5 = ({}^\ell\delta_{ix})\{{}^\ell\delta_{jz}\,{}^\ell\delta_{ky}\left|\frac{I_{xx\ell}+I_{yy\ell}-I_{zz\ell}}{2}\right| + |{}^\ell\delta_{jz}\,{}^\ell d_{kx}-{}^\ell\delta_{jx}\,{}^\ell d_{kz}|$$
$$m_\ell z_\ell\}$$

$$Q_6 = \{{}^\ell\delta_{jx}\,{}^\ell\delta_{ky}x_\ell+|{}^\ell\delta_{jz}(-{}^\ell\delta_{kz})-{}^\ell\delta_{jx}\,{}^\ell\delta_{kx}|y_\ell+{}^\ell\delta_{jz}\,{}^\ell\delta_{ky}\,z_\ell+$$
$$|{}^\ell\delta_{jz}\,{}^\ell d_{kx}-{}^\ell\delta_{jx}\,{}^\ell d_{kz}|\}^\ell d_{iy}m_\ell$$

$$m_{33} = Q_7 + Q_8 + Q_9$$

where

$$Q_7 = ({}^\ell\delta_{iy})\{{}^\ell\delta_{kz}{}^\ell\delta_{jx}\left|\frac{-I_{xx\ell}+I_{yy\ell}+I_{zz\ell}}{2}\right| + (-{}^\ell\delta_{jy}){}^\ell d_{kx} + {}^\ell\delta_{jx}{}^\ell d_{ky}\right| \\ m_\ell x_\ell\}$$

$$Q_8 = ({}^\ell\delta_{ix})\{{}^\ell\delta_{kz}{}^\ell\delta_{jy}\left|\frac{I_{xx\ell}-I_{yy\ell}+I_{zz\ell}}{2}\right| + (-{}^\ell\delta_{jy}){}^\ell d_{kx} + {}^\ell\delta_{jx}{}^\ell d_{ky}\right| \\ m_\ell y_\ell\}$$

$$Q_9 = \{{}^\ell\delta_{jx}{}^\ell\delta_{kz}x_\ell + {}^\ell\delta_{jy}{}^\ell\delta_{kz}y_\ell + |(-{}^\ell\delta_{jy}){}^\ell\delta_{ky} + {}^\ell\delta_{jx}(-{}^\ell\delta_{kx})|z_\ell + \\ |(-{}^\ell\delta_{jy}){}^\ell d_{kx} + {}^\ell\delta_{jx}{}^\ell d_{ky}|\}{}^\ell d_{iz}m_\ell$$

Simulation Results:

Computer programs were written to validate the simplified formulation by comparing the force and torque values calculated by

(a) the classical Lagrangian formulation and
(b) the simplified formulation developed in this paper.

For our example a stanford manipulator was simulated.

As an example, each joint position parameter is chosen as $\Theta_i = \emptyset.3$ radians (i=1,2,...,6) for all the cases considered in our example. The resulting joint forces and torques were calculated for the different joint velocities and accelerations. The velocities and accelerations were chosen to give realistic simulation results, whilst maintaining consistency with the existing robot models.

Case (1):

joint velocities $(\dot{\Theta}_i) = 0.5$ rad/S.
joint accelerations $(\ddot{\Theta}_i) = 0.5$ rad/S^2.

Table 1.

Joint / Case	1(Nm)	2(Nm)	3(N)	4(Nm)	5(Nm)	6(nm)
I	0.754	3.96	-57.26	0.0007	0.656	0.0105
II	0.739	4.041	-57.38	0.0074	0.7186	0.01048

I: Lagrangian

II: This paper model.

Case (2):

joint velocities $(\dot\theta_i)$ = 0.8 rad/S.
joint accelerations $(\ddot\theta_i)$ = 0.8 rad/S^2.

Table 2.

Joint Case	1(Nm)	2(Nm)	3(N)	4(Nm)	5(Nm)	6(Nm)
I	1.1	4.91	-55.48	0.017	0.706	0.0168
II	0.946	4.95	-55.96	0.076	0.84	0.017

Case (3):

joint velocities $(\dot\theta_i)$ = 1 rad/S.
joint accelerations $(\ddot\theta_i)$ = 1 rad/S^2.

Table 3.

Joint Case	1(Nm)	2(Nm)	3(N)	4(Nm)	5(Nm)	6(Nm)
I	1.178	5.432	-54.47	0.026	0.74	0.0211
II	1.184	5.604	-55.07	0.11	0.95	0.0209

Case (4):

joint velocities $(\dot\theta_i)$ = 1.5 rad/S.
joint accelerations $(\ddot\theta_i)$ = 1.5 rad/S^2.

Table 4.

Joint Case	1(Nm)	2(Nm)	3(N)	4(Nm)	5(Nm)	6(Nm)
I	0.63	6.18	-52.79	0.05	0.82	0.0318
II	0.392	7.17	-53.3	0.0686	0.788	0.0313

The previous four cases were plotted to show the difference between the two schemes, as shown in Fig.1.

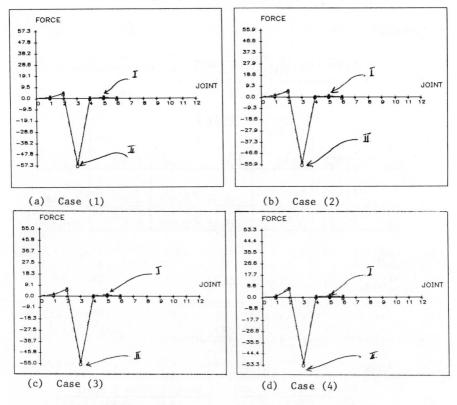

(a) Case (1)   (b) Case (2)

(c) Case (3)   (d) Case (4)

Fig.1 Forces and torques effecting each joint in case
       (I) and (II).

In calculating the forces and torques effecting each joint,
some simplifications were used to increase the speed and
efficiency of the calculations $|3,7|$ such as $P_{...i}=0$ always,
because the joint which generates the centripetal force will
not be effected by it, and $P_{ijk}=0$, if i=K, K>j.

Concluding Remarks:

The equations which describe the behaviour of a robot manipulator consist of a considerable number of differential
and non-linear terms which complicate the calculations and
make the real-time control a very hard task. Coriolis and
Centripetal effects are of special importance in high speed
movements. These effects were studied and a simplified representation was produced based on the Lagrangian formulation.
Simulation and numerical results were presented at different
speeds (see table 1,2,3,4) for comparison between the classical Lagrangian formulation of coriolis and centripetal
effects and the simplified model.

References:

(1)   Paul, R,P, 1981, "Robot Manipulators: Mathematics,
        Programming and Control", MIT Press, Cambridge,Mass.

(2)   Paul, R,P, 1972, "Modelling, Trajectory Calculation
        and Servoing of a Computer Controlled Arm", Stanford
        Artificial Intelligence Laboratory, Stanford
        University, AIM 177.

(3)   Bejczy,A,K, 1974, "Robot Arm Dynamics and Control",
        NASA-JPL Technical Memorandum, 33-669.

(4)   Hollerbach,J,M, 1980, "A Recursive Lagrangian Formul-
        ation of Manipulator Dynamics and a Comparative Study
        of Dynamics Formulation", IEEE Trans. on Systems, Man,
        and Cybernetics, SMC.10, pp 730-736.

(5)   Lee,C.S. et al., 1983, "Development of the Generalized
        D'Alembert Equations of Motion for Mechanical Manipul-
        ators". Proc. of 22nd Conf. on Decision and Control,
        Dec.14-16.

(6)   Orin D,E, et al., 1979, "Kinematic and Kinetic Analy-
        sis of Open Chain Linkages Utilizing Newton-Euler
        Methods", Math.Biosc., 43, 107-130.

(7)   Liegeois,A, 1984, "Performance and Computer-Aided
        Design", Robot Technology, Volume 7.

(8)   Luh, J,Y,S, et al., 1980, "On-line Computational
        Scheme for Mechanical Manipulators", Trans. ASME, J.
        Dynamic Sytems, Meas., and Control, 102, pp 69-76.

(9)   Walker, M.W., Orin,D,E, 1982, "Efficient Dynamic Com-
        puter Simulation of Robotic Mechanisms", Trans. ASME,
        J Dynamic Systems, Meas., and Control, 104, pp 205-211.

(10)   Denavit,H., Hartenberg,R, 1955 "A Kinematic Notation
        for Lower Pair Mechanisms Based on Matrices", J.
        Applied Mechanics, 22, pp 215-221.

(11)   Silver,W,M., 1982, "On the Equivalence of Lagrangian
        and Newton-Euler Dynamics for Manipulators", Int.J.
        of Robotics Res., Vol.1, No.2.

(12)   Raibert,M.H., Horn, B.K, 1978, "Manipulator Control
        using the Configuration Space Method", The Industrial
        Robot, Vol.5, No.2, pp 69-73.

(13)   Kane,T., Levinson,D, 1983, "The Use of Kane's Dynami-
        cal Equations in Robotics", Int.J.Robotics Res., Vol.2,
        No.3.

(14)   Mahil S,S, 1982, "On the Application of Lagrange's
        Method to the Description of Dynamic Systems", IEEE
        Trans, Systems, Man and Cybernetics, Vol.SMC.12,
        pp 877-889.

(15)   Armstrong, W,W, 1979, "Recursive Solution to the
        Equations of Motion of an N-link Manipulator", Proc.
        5th World Congress on Theory of Machines and Mechan-
        isms.

(16)   Lee., C,S, Chang, P,R, 1986, "Efficient Parallel
        Algorithm for Robot Inverse Dynamics Computation",
        IEEE Trans. Systems, Man and Cybernetics, Vol.SMC.16,
        No.4.

(17)   Binder., E,E, Herzog, J,H, 1986, "Distributed
        Computer Architecture and Fast Parallel Algorithms in
        Real-Time Robot Control", IEEE Trans. Systems, Man
        and Cybernetics, Vol.SMC.16, No.4.

(18)   Luh, J,Y,S., Lin, C,S, 1982, "Scheduling of Parallel
        Computation for a Computer Controlled Mechanical
        Manipulator", IEEE Trans. Syst, Man and Cybernetics.
        Vol.SMC.12, pp 214-234.

(19)   Lathrop, L,H, 1983, "Parallelism in Manipulator
        Dynamics", MIT Artificial Intelligence Lab.,
        Cambridge, MA, Tech.Rep.No.754.

Chapter 9

Robot manipulator trajectory tracking and parameter estimation

M. Dickinson and A. S. Morris

1   INTRODUCTION

This paper highlights the need for suitable sensors to assess robot manipulator dynamic performance. Tracking techniques are briefly discussed and their possible application to manipulator control.

The paper considers in some detail the operation of a real-time tracking system and shows how this may be used for parameter estimation of robot joint drives. The estimated parameters may be used to determine better controller terms or to modify the dynamic model of the manipulator, for position control or computed torque control schemes respectively.

2   ROBOT MANIPULATOR TRAJECTORY TRACKING

Robot manipulators have traditionally been controlled using individual three-term controllers for each joint. The reference input for each joint is computed by a supervisory computer from a defined trajectory and the manipulator kinematic model. Accuracy of end-effector positioning is therefore dependent upon the ability of each joint servo to maintain the desired angle under varying inertial and frictional load, and upon the accuracy of the kinematic model.

Recent research has led to the development of the "computed torque" technique, where the torque requirement for each joint is computed from a defined trajectory and the manipulator dynamic model as used by Luh et al (1), (2), Koivo (3), Norcross et al (4) and Chung and Cho (5). This scheme produces much improved control of the manipulator in terms of trajectory adhesion, but is excessive in terms of computing requirements, since the manipulator dynamics must be computed in real time. Accuracy of end effector positioning in this case is dependent upon the accuracy of the manipulator kinematic and dynamic models, although joint position feedback is often added to ensure good positioning under varying load conditions and where unmodelled effects cause deviation from the desired trajectory.

Both of the above control schemes suffer from the same deficiencies, in that they are effectively 'open loop' as far as the manipulator itself is concerned. There is no feedback from the end effector relative to its actual position in Cartesian space. Several researchers have attempted to reduce the Cartesian error, including Backers et al (6) and Lim and Eslami (7) . All these schemes lack appropriate feedback since suitable sensors have not until recently existed.

Manipulator trajectory tracking has become an area of increasing interest for several reasons. Robot calibration and performance assessment has been carried out using static measurement techniques, but users are demanding a performance indicator of trajectory adhesion capability as robots are applied to more intricate assembly tasks. Coupled with this need is the requirement to assess robot velocity and acceleration in the work space.

The availability of such information in real time allows the synthesis of new control forms for robot manipulators. It is anticipated that the application of a real time tracking sensor would effectively close the loop and realise benefits in terms of trajectory adhesion and performance.

3   MANIPULATOR TRACKING SYSTEMS

Measurement of the x, y, z coordinates of the manipulator end-effector have been attempted using various techniques, including theodolite, 3 wire triangulation, ultrasonic and laser tracking techniques. The most notable include Chande and Sharma (8), Moritz et al (9), Dunkin and Thaler (10), and in particular the laser tracking systems of Gilby and Parker (11) and Lau et al (12). The only tracking system suitable for use as a feedback sensor is the one described by Dickinson and Morris (13) since it produces results in real time.

The principle of operation of these systems is generally shown in Figure 1, where the spatial location of a point x, y, z, is determined from two fixed known distances d, e, and measurement of three other distances a, b, c, from

$$x = \frac{a^2 + d^2 - b^2}{2d} \tag{3.1}$$

$$y = \frac{a^2 + e^2 - c^2}{2e} \tag{3.2}$$

$$z = \sqrt{a^2 - x^2 + y^2} \tag{3.3}$$

This is implemented in the Dickinson and Morris system as shown in Figure 2.

Distances a, b and c are measured by timing the propagation of a burst of ultrasonic energy from the transmitter mounted on the end-effector to three receivers orthogonally placed at S_1, S_2 and S_3. A microcomputer system is used to compute (3.1), (3.2) and (3.3), producing the x, y, z coordinates. The velocity in each direction is then computed by numerical differentiation. The sensor therefore produces a time sequence of coordinates and velocities, in real-time. This data can be saved on disc for later offline analysis. The system can also be programmed to compute the manipulator inverse kinematics, therefore producing the joint space trajectories, as shown in Figure 3.

As far as each joint is concerned, there is then available an input-output sequence which may be used for parameter estimation. Use of the output sequence from the tracking system is justified since it takes into account the effects of link flexure, backlash and other unmodelled effects. Thus the data may be used to ascertain the manipulator kinematic and dynamic models.

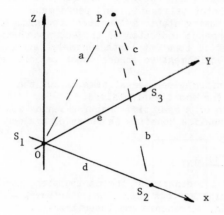

Figure 1   Spatial location of a point, P

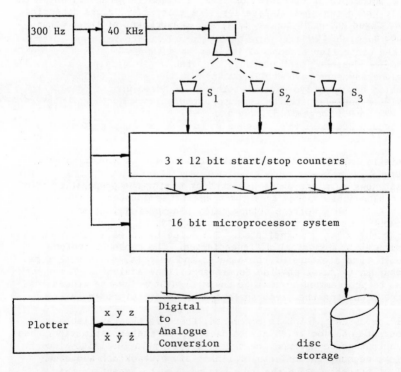

Figure 2   Tracking System Configuration

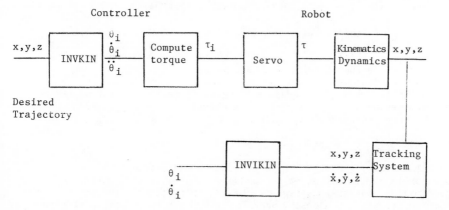

Figure 3   Joint space information from tracking

4   PARAMETER ESTIMATION

As the manipulator configuration changes, the dynamic performance changes and hence the parameters of the individual joint drives also change. Knowledge of the joint parameters leads to improved control since the controller or control law may be suitably modified to produce the desired joint response over the whole working envelope. Several researchers have published work on the subject of parameter estimation and adaptive control techniques, based upon models of the manipulator and evaluated by computer simulation. Worthy of note are Dubowsky and Desforges (14), Lee and Lee (15), Anex and Hubbard (16), Horowitz and Tomizuka (17), Backes et al (18) and Craig et al (19).

The real-time tracking system (13) may be used to provide the output sequence. Dickinson and Morris (20) describe how the input-output sequence may be used in a recursive least squares algorithm to estimate joint inertia and friction in a torque drive.

The joint is modelled as a simple second order system,

$$\begin{bmatrix} \overset{o}{x}_1 \\ \overset{o}{x}_2 \end{bmatrix} = \begin{bmatrix} A_{11} & A_{12} \\ 0 & A_{22} \end{bmatrix} \begin{bmatrix} x_1 \\ x_2 \end{bmatrix} + \begin{bmatrix} 0 \\ B_{21} \end{bmatrix} V_i \qquad (4.1)$$

where x_1 and x_2 are motor speed and torque respectively, (4) is rewritten in discrete form

$$\omega_{(z)} = [z^{-1}\omega \quad z^{-2}\omega \quad z^{-1}V_i] \begin{bmatrix} \theta_1 \\ \theta_2 \\ \theta_3 \end{bmatrix} \qquad (4.2)$$

where θ is the parameter vector to be estimated.

Many algorithms have been proposed for parameter estimation of such systems, but for real-time completion a least squares approach seems efficient, the algorithm

$$\theta(k) = \theta(k-1) - P(k)z(k)[Z^T(k)\theta(k-1) - y(k)] \tag{4.3}$$

$$\text{where, } P(k) = P(k-1) - p(k-1)z(k)z^T(k)P(k-1)[1+Z^T(k)P(k-1)z(k)]^{-1} \tag{4.4}$$

is proposed, which may easily be computed within the time available on-line, since there are no matrix inversions,

Since joint parameters vary with arm configuration, tracking capability is essential, hence a "forgetting factor" is introduced to adjust the speed at which adaptation occurs.
(Taken from "Computer Control of Industrial Processes" (21))

5   APPLICATION AND RESULTS

The tracking system may be applied to conventional position controlled manipulators, or to torque controlled manipulators. For performance assessment purposes, the manner of joint drive is immaterial, since trajectory adhesion and accuracy of positioning is of greater importance. For parameter estimation, both systems may be represented in standard second order form,

$$\frac{\theta_o}{\theta_i} = \frac{k\,\omega_n^2}{s^2 + 2\zeta\omega_n s + \omega_n^2} \tag{5.1}$$

and hence the discrete form

$$\theta_o(z) = [z^{-1}\theta_o \quad z^{-2}\theta_o \quad z^{-1}\theta_i] \begin{bmatrix} \theta_1 \\ \theta_2 \\ \theta_3 \end{bmatrix} \tag{5.2}$$

Choice of states however, demands attention since the tracking system produces joint angle and angular velocity.

Figure 4 shows the results of tracking a PUMA model 560 manipulator, as the end-effector described a horizontal square (ie in (x, y)). Note how x-y motion may be analysed, as well as x-x or y-y for velocity and acceleration analysis.

Dickinson and Morris in (20) describe results of parameter estimation. The parameters were accurate to 14% after 100 samples, and 4% after 200 samples.

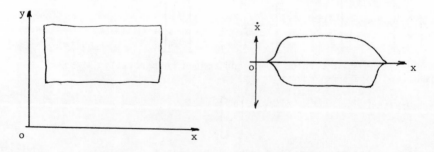

Figure 4   Tracking system results

6   CONCLUSION

A robot manipulator tracking system has been described, capable
of analysing Cartesian and joint space trajectories.   Results have
shown that velocity and acceleration profiles may be obtained.

The tracking system has been applied to the manipulator parameter
estimation problem.   Initial results from computer simulations have
shown that this is feasible for position or torque controlled
manipulators.

This work is to continue to identify an industrial robot, for
verification of dynamic and kinematic models.   It is anticipated that
the scheme may be extended to allow measurement of link flexure or
joint backlash.

REFERENCES

(1) Luh J Y S, Walker M W, and Paul R P C
     "On-line Computational Scheme for Mechanical Manipulators"
     Journal of Dyn. Sys. Meas. and Control, 102, 69-76

(2) Luh J Y S, Walker M W, and Paul R P C
     "Resolved Acceleration Control of Mechanical Manipulators"
     IEEE Trans. Automatic Control, 25 468-474

(3) Koivo A J
     "Self-tuning Manipulator Control in Cartesian Base
     Coordinate System". Trans. of the ASME, 107, 316-323

(4) Norcross R J, Wang J C, McInnis B C and Shieh L S
     "Pole Placement Methods for Multivariable Control of Robotic
     Manipulators". Trans. of the ASME, 108, 340-345

(5) Chung W K and Cho H S
     "on the Dynamics and Control of Robotic Manipulators with an
     Automatic Balancing Mechanism". Procs. I.Mech.E, 201, 25-34

(6) Backes P G, Leininger G G and Chun-Hsien Chung
     "Joint Self-Tuning with Cartesian Setpoints". Trans. of the
     ASME, 108, 146-150

(7) Kye Y Lim and Mansour Eslami
     "Adaptive Controller Designs for Robot Manipulator Systems
     Yielding Reduced Cartesian Error". IEEE Trans. Automatic Control
     ACC-32, 184-187

(8) Chande P K and Sharma P C
     "A Fully Compensated Digital Ultrasonic Sensor for Distance
     Measurement". IEE Trans. on Inst. and Meas., IM33, 128-129

(9) Moritz W E, Shreve P L and Mace L E
     "Analysis of an Ultrasonic Spatial Locating System". IEE
     Transactions on Inst. and Meas., 25, 43-50

(10) Dunkin W M and Thaler C J
     "Ultrasonic Position Sensing for a Sentry RObot and/or a
     Robot Manipulator Arm". Second International Conference on
     Real-Time Control of Electromechanical Systems, 1986, IERE

(11) Gilby J H and Parker G A
     "Robot Arm Position Measurement Using Laser Tracking
     Techniques". Proceedings of the 7th British Robot
     Association Annual Conference, 1984.

(12) Lau K, Hocken R and Haynes L
     "Robot Performance Measurements using Automatic Laser Tracking
     Techniques". Robotics and Comp.Int. Manuf., 2 No 3/4,
     227-236

(13) Dickinson M and Morris A S
"Coordinate Determination and Performance Analysis for Robot
Manipulators and Guided Vehicles". Accepted for publication
in IEE Procs Pt A.

(14) Dubowsky S and DesForges D T G
"The Application of Model-Referenced Adaptive Control to
Rototic Manipulators". Journal of Dyn. Sys., Meas. and Control
101, 193-200

(15) Lee C S G and Lee B H
"Resolved Motion Adaptive Control for Mechanical
Manipulators". Trans. of the ASME, 106, 134-142

(16) Anex R P and Hubbard M
"Modelling and Adoptive Control of a Mechanical
Manipulator". Journal of Dyn. Sys., Meas. and Control, 106,
211-217

(17) Horowitz R and Tomizuka M
"An Adaptive Control System for Mechanical Manipulators -
Compensation of Non-linearity and Decoupling Control".
Journal of Dyn. Sys., Meas. and Control, 108, 127-135

(18) Backes P G, Leininger G G and Chun-Hsien Chung
"Joint Self-Tuning with Cartesian Setpoints". Trans. of the
ASME, 108, 146-150

(19) Craig J J, Psing Hu and Shankar Sastry
"Adaptive Control of Mechanical Manipulators". IEEE
Conference "Robotics and Automation" San Fransisco April 1986

(20) Dickinson M and Morris A S
"A Torque Drive for Robot Manipulators with Load Parameter
Estimation". Submitted to IEE Procs Pt D

(21) Bennett S and Linkens D A (Edited)
"Computer Control of Industrial Processes". Peter Peregrinus

Chapter 10

A decentralised high gain controller for trajectory control of robot manipulators

S. R. Pandian, M. Hanmandlu and M. Gopal

ABSTRACT

A new decentralised high gain feedback algorithm is proposed for the trajectory control of robot manipulators. Model
following controllers are designed for joint actuator models
by treating the influence of the manipulator dynamics as a
disturbance. The proposed scheme is simple, computationally
easy and robust to parametric uncertainities and payload
variations.

1. INTRODUCTION

The model reference adaptive control (MRAC) approach,
also known as adaptive model following control, provides a
powerful tool in the control of complex, uncertain dynamical
systems such as robot manipulators (Dubowsky and Desforges
[1], Young [2] and Sundareshan and Koenig [3]). The desired
performance of the nonlinear, time-variant system, the 'plant'
is specified to be the behaviour of a linear, time-invariant
system, the 'model'. The control problem is thus reduced to
controlling the dynamics of the error between the plant and
model states.

In interconnected systems control, a well-known approach
is to consider the interactions between subsystems as disturbances and reject them by disturbance rejection methods
(Utkin [4]). In the present paper, this approach is refor-
mulated in a model following framework, by treating the stable (or locally stabilised) isolated subsystem as a 'model'
for the interconnected subsystem, so that the stabilisation
of the latter becomes a problem of model following.

Incorporation of the model following behaviour requires
an adaptive control or a non-adaptive control such as the
variable structure or high gain feedback. We have treated
the variable structure systems approach elsewhere (Pandian
et al.[5]) and here the high gain control methodology is
taken up for the trajectory control of robot manipulators.

2. THE PROPOSED CONTROLLER

The problem of decentralised control of robot manipulators is simplified by dealing with the dynamic models of
joint actuators, alongwith that of the manipulator (Stokic

and Vukobratovic [6]).

We consider the dynamics of the i-th joint actuator of an N-link manipulator represented as a linear, time-invariant system

$$\dot{x}_i = A_i x_i + b_i u_i + d_i f_i \qquad (1)$$

where the state x_i is $n_i \times 1$ and control u_i is a scalar Neglecting the actuator friction torque, f_i is the driving torque acting on the actuator and is given by the nonlinear, uncertain dynamics of the manipulator :

$$f_i = H_i(q,d) \ddot{q} + h_i(q,\dot{q},d) \qquad (2)$$

Here, $q = (q_1, q_2, \ldots, q_N)'$ is the N-vector of joint angles or displacements and d is an ℓ-vector of parameters (such as payload and link masses and inertias). H_i is the N-vector of inertia and h_i represents Coriolis, centrifugal and gravity forces.

In each x_i, two coordinates coincide with q_i and $\dot{q}_i$. In the general case, q_i and $\dot{q}_i$ are nonlinear functions of x_i and u_i is amplitude-constrained.

In the robot trajectory tracking problem, the system must track the desired or nominal trajectories $(q_{r1}(t), q_{r2}(t), \ldots, q_{rN}(t))$ over a time interval of interest $[0,T]$. The nominal velocity trajectories $(\dot{q}_{r1}, \dot{q}_{r2}, \ldots, \dot{q}_{rN})$ are readily obtained. We also need to specify the remaining variables in each x_i. Let x_{ri} be the corresponding nominal state trajectory.

We assume that there exists a nominal control $u_{ri}(t)$, $t \in [0,T]$, which forces the 'isolated' subsystem in (1) to track the nominal state trajectory, i.e.,

$$\dot{x}_{ri}(t) = A_i x_{ri}(t) + b_i u_{ri}(t) \qquad (3)$$

The state tracking error is $e_i = x_i - x_{ri}$.

From (1) and (3), the error dynamics is

$$\dot{e}_i = A_i e_i + b_i \tilde{u}_i + d_i f_i \qquad (4)$$

where $\tilde{u}_i = u_i - u_{ri}$.

The control $u_i = u_{ri} + \tilde{u}_i$ thus consists of two components : the nominal control u_{ri} which helps track the nominal trajectories in the absence of the uncertain dynamics, and a compensating control $\tilde{u}_i$ which neutralises the effects of non-zero initial error $e_i(0)$ and the effects of interlink coupling, Coriolis, centrifugal and gravity forces and of payload variations.

From (4), we see that the tracking problem is solved by choosing a high gain control law for u_i such that e_i tends to zero asymptotically. Since we can rewrite (4) as

$$\dot{e}_i = A_i \, e_i + b_i \, \tilde{u}_i + d_i \, f_i(x) \tag{5}$$

the subsystem error dynamics depends on the evolution of the interconnected system state $x'=(x_1', x_2', \ldots, x_N')$. For implementation of local feedback, we can treat (5) as a centralised system by considering $f_i(x)$ as a disturbance, representing the complexities posed by the manipulator. So, to ensure the robustness of the controller, invariance of error state e_i to the disturbance f_i is necessary.

The condition for invariance of the state vector with respect to the disturbance is given by

$$\text{rank}(b_i) = \text{rank } (b_i \; \vdots \; d_i) \tag{6}$$

For the actuator in (1), this condition, in general, does not hold good. However, a less stringent condition applies if invariance only of outputs is required.

Let the output be

$$y_i = C_i \, e_i = [C_1 \; \vdots \; C_2] \begin{bmatrix} x_1 \\ x_2 \end{bmatrix} \tag{7}$$

where the system (4) with high gain control

$$\tilde{u}_i = \frac{1}{\varepsilon_i} (K_2 \, K_1 \, x_1 + K_2 \, x_2) \tag{8}$$

is transformed as

$$\dot{x}_1 = A_{11} \, x_1 + A_{12} \, x_2 + d_1 \, f \tag{9}$$

$$\dot{x}_2 = (\varepsilon \, A_{21} + b_2 \, K_2 \, K_1) x_1 + (\varepsilon \, A_{22} + b_2 \, K_2) x_2 + d_2 \, f \tag{10}$$

where subscript i has been omitted for simplicity, and $g_i = 1/\varepsilon_i$ is a large scalar gain factor.

The condition for asymptotic disturbance decoupling is (Young [7])

$$\langle A_{11} - A_{12} \, K_1 \mid \mathcal{R}(d_1) \rangle \subseteq \mathcal{N}(C_1 - C_2 \, K_1) \tag{11}$$

where $\mathcal{R}$ and $\mathcal{N}$ are range space and null space respectively.

While (11) is a less stringent condition than (6), it is not always satisfied for the problem under consideration. An example is the case when $(A_{11} - A_{12} \, K_1, \; d_1)$ is a controllable pair, and the position tracking error is treated as the output.

Alternatively, system (4) can be decomposed into n_i first-order systems, using transformations of the type (9)-(10). The conditions and control gains for invariance can be obtained by applying composite Lyapunov function techniques (Utkin et al. [8]).

In this paper, we consider the high gain control law

$$\tilde{u}_i = g_i \, k_i^! \, e_i \tag{12}$$

where k_i : $n_i \times 1$ is the subsystem local gain vector, and to develop a procedure that ensures disturbance decoupling as well as provides estimates of minimum gain factors required for specified disturbance bounds.

To achieve disturbance decoupling, we introduce a change in coordinates, transforming the system (4) to a form that satisfies condition (6). For this, we identify a coordinate y which contains data on the magnitude of the (generally un-measurable) disturbances (Utkin [9]). For example, in a dc motor-driven arm, with state $x_i^! = (q_i, \dot{q}_i, i_a)$, force/torque sensing is done simply by measuring the armature current (Fu et al. [10]).

Rewriting (4) in the form

$$\dot{x}^* = A^* \, x^* + b^* \, u + d^* \, f \tag{13}$$

Let y be in one of the first $(n-1)$ equations of (13) and this equation, say k-th independent of control :

$$\dot{x}_k = \sum_{i=1}^{n-1} a_{ki}^* \, x_i + a_{kn}^* \, y + d_k^* f \tag{14}$$

We introduce the new variable $x_n = \dot{x}_k$, and eliminate y from the system of equations (13), by using eq.(14).

If $x' = (x_1, x_2, \ldots, x_n)$ is the new state, then (13) is transformed into

$$\dot{x} = Ax + bu + df + d_1 \dot{f} \tag{15}$$

where the coefficients of A, b, d and d_1 depend on the coefficients of (13) and (14).

In the single-input system (4), the control enters only in one equation, while the scalar disturbance enters only in the equation for acceleration. This implies that the vectors in (15) are of the form

$$b' = (0 \, \ldots \, 0 \, b_n), \; d' = (0 \, \ldots \, 0 \, d_n), \; d_1^! = (0 \, \ldots \, 0 \, d_{1n}).$$

The invariance condition (6) is, therefore, satisfied. To obtain estimates of g_i as a function of interconnections, we manipulate (15) as below

$$\dot{x} = Ax + bu + df + d_1 \dot{f}$$
$$= Ax + bu + d(f + D \, d_{1n} \, \dot{f}) \; ; \; D = 1/d_n$$
$$= Ax + b(u + B \, d_n(f + D \, d_{1n} \, \dot{f}) \; ; \; B = 1/b_n$$

or

$$\dot{x} = Ax + b(u+F) \tag{16}$$

where $F = B \, d_n(f + D \, d_{1n} \, \dot{f})$.

The decentralised version of (16) is

$$\dot{\tilde{e}}_i = \tilde{A}_i \, \tilde{e}_i + \tilde{b}_i \, (\, \tilde{u}_i + F_i) \tag{17}$$

where $\tilde{A}_i$, $\tilde{b}_i$ and $\tilde{e}_i$ are transformed matrices and vectors, and $\tilde{u}_i$ is the high gain control.

For a more general case of (17)

$$\dot{x}_i = f_i(x_i) + b_i(u_i + g_i(x)) \tag{18}$$

upper bounds on $\varepsilon_i = 1/g_i$ have been derived by Saberi and Khalil [11], using the composite Lyapunov function method.

We assume $f_i(x)$ and $\dot{f}_i(x)$ to be bounded disturbances

$$|f_i| \leq F_{1i} \quad \text{and} \quad |\dot{f}_i| \leq F_{2i} \tag{19}$$

where F_{1i} and F_{2i} may be estimated as bounds on the nominal torque trajectories for a given tracking task. F_{1i} may also be taken as the maximum torque deliverable by the i-th actuator, but this gives more conservative gains.

These estimates are used in the calculation of bounds on ε_i, assuming bounds on error variables, $|e_{ij}|$. The latter, in turn, can be minimised by minimising the initial errors $e_{ij}(0)$.

3. AN EXAMPLE

To illustrate the application of the proposed algorithm to robot trajectory control, we consider the UMS-2 cylindrical manipulator, in minimal configuration (N=3) discussed by Stokic and Vukobratovic [6].

The joints are driven by dc motor actuators, modelled as third order systems. A_i's are constant matrices, $b_i' = (0 \; 0 \; b_{i3})$, and $d_i' = (0 \; d_{i2} \; 0)$, so invariance conditions do not hold good.

Since finite high gain control is essentially a proportional feedback, integral feedback of position tracking error is used to eliminate steady state error. The control law then becomes

$$\tilde{u}_i = g_i \, [k_{pi}' \, e_i + k_{Ii} \int_0^t (q_i - q_{ri}) d\tau] \tag{20}$$

k_{pi} and k_{Ii} are chosen by solving a pole placement problem for the slow and fast subsystems [12] of (4), with e_i augmented by $z_i = \int_0^t (q_i - q_{ri}) d\tau$ as an additional state variable. g_i are chosen sufficiently large. Alternatively, lower bounds on g_i can be estimated for specified disturbance bounds for the augmented system.

To examine the robustness of the proposed controller, the nominal trajectories are specified as shown in Fig.(1). Initial errors are chosen as $e_{i1} = (0.1 \quad 0.01 \quad 0.01)$, $e_{i2}= \{0\} = e_{i3}$. The nominal payload mass and inertia are (5.6 kg, 0.32 kg-m^2) with off-nominal values (9.6, 1.0). The proportional and integral gains used are listed in Table 1.

The simulation results are shown in Figs.(2) to (6). As seen from Figs.(2) to (4), the tracking is accurate and robust to payload variations. The torque and control voltage trajectories have been shown in Figs.(5) and (6)for the case of link-2.

Table-1

Link	g_i	k_{pi}	k_{Ii}
1	2000	$-(0.0581, 0.00825, 0.000375)$	-0.13125
2	250	$-(3.75, \quad 0.5625, \quad 0.02679)$	-8.0356
3	100	$-(6.4615, 0.9692, \quad 0.04615)$	-13.846

4. DISCUSSION

The advantages of the proposed controller are :

(i) It is based on the relatively simple actuator models, whereas a majority of the existing robot control algorithms are based on the complex manipulator model. Extension of our scheme to the case of non-linear, uncertain actuator dynamics is fairly straightforward.

(ii) The feedback is decentralised. This leads to a simple controller structure and a considerable reduction in computations.

(iii)The non-adaptive controller proposed here - like its VSS counterpart in [5] - retains the robustness properties of the adaptive controllers without much computational requirements.

(iv) Electrical, hydraulic and pneumatic actuators can be modelled as single-input systems, treating the driving torque as a scalar disturbance. So they can all be represented by eq.(1), which in turn can be transformed into (15) for invariant system design. If lower order actuator models are used to reduce computations (Vukobratovic et al. [13]), the system may directly satisfy the invariance condition. In such cases, transformation to the form (15) is not required.

To eliminate 'chattering' in variable structure systems, a high gain controller replaces the discontinuous controller in the vicity of the so-called 'sliding mode' [2]. So the results of this paper are directly relevant to the decentralised VSS scheme in [5].

The concept of a 'nominal system' with a nominal control has been employed in many papers on trajectory control (e.g. [6], [13]-[16]). By treating this nominal system as a 'model', the controller structure has been simplified in our case. For example,the controller in [6] requires additional global force feedback or adapative feedback,if local feedback is found insufficient. Similarly, the controller in [14] is fairly complex. The schemes in [15] and [16]are based   on high gain feedback, and are centralised. The control law is more complicated than ours, and requires nominal torque calculations, using inverse dynamics.

5. CONCLUSION

By dealing with the actuater models, and treating   the influence of manipulator complexity as a disturbance effect, a new decentralised trajectory control algorithm based   on high gain error feedback has been proposed. The controller has a simple structure, is computationally easy to design, and is robust to manipulator uncertainties and payload variations.

6. REFERENCES

1. Dubovsky, S., and DesForges, D.T., 1979, ASME J.Dyn.Syst. Meas. and Control, 101, 193-200.

2. Young, K.D., 1986, 'A Variable Structure Model Following Control for Robotic Applications,' Proc. IEEE Int. Conf. Robotics and Automation, San Francisco, 540-545.

3. Sundareshan, M.K., and Koenig, M.A., 1985, 'Decentralised Model Reference Adaptive Control of Robotic Manipulators,' Proc. American Cont. Conf., Boston, 44-49.

4. Utkin, V.I., 1977,'Variable Structure Systems with Sliding Modes : A Survey,' IEEE Trans. Auto.Cont., AC-22, 212-222.

5. Pandian, S.R., Hanmandlu, M., and Gopal, M., 1987, 'A Decentralised Variable Structure Model Following Controller for Robot Manipulators,' submitted to the 1988 IEEE Int.Conf. Robotics and Automation, Philadelphia.

6. Stokic, D., and Vukobratovic, M., 1984, Automatica, 20, 353-358.

7. Young, K.D., 1982, IEEE Trans. Auto.Cont., AC-27, 970-971.

8. Utkin, V.I., Drakunov, S.V., Izosimov, D.E., Lukyanov,A.G. and Utkin, V.A., 1984, 'A Hierarchical Principle of the Control System Decomposition based on Motion Separation,' Proc.IFAC 9th Triennial World Congress, Budapest,1553-1558

9. Utkin, V.I., 1978, 'Sliding Modes and their Application in Variable Structure Systems,' Mir, Moscow.

10. Fu, K.S., Gonzalez, R.C., and Lee, C.S.G.,1987, 'Robotics: Control, Sensing, Vision and Intelligence,'McGraw-Hill, New York.

11. Saberi, A., and Khalil, H., 1982, Int.J.Control, 36, 803-818.

12. Young, K.D., Kokotovic, P.V. and Utkin, V.I., 1977, IEEE Trans. Auto. Cont., AC-22, 931-938.

13. Vukobratovic, M., Stokic, D., and Kircanski, N., 1985, 'Non-Adaptive and Adaptive Control of Manipulation Robots,' Springer Verlag, Berlin.

14. Shoureshi, R., Corless, M.J., and Roesler, M.D., 1987, ASME J. Dyn. Syst. Meas. and Control, 109, 53-59.

15. Mills, J.K., and Goldenberg, A.A., 1986, 'A New Robust Robot Controller,' Proc. IEEE Int. Conf. Robotics and Automation, San Francisco, 740-745.

16. Mills, J.K., and Goldenberg, A.A., 1986, 'Robust Control of a Robotic Manipulator with Joint Variable Feedback,' Proc. ASME Winter Meeting on 'Robotics : Theory and Applications,' 43-49.

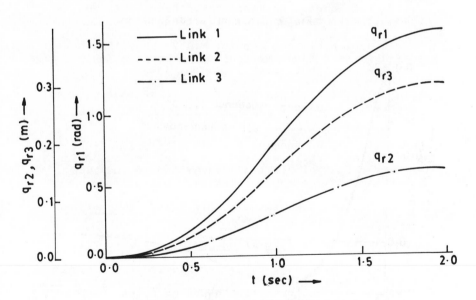

Fig. 1   Nominal trajectories

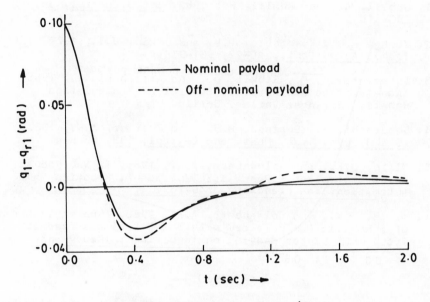

Fig. 2 Tracking error : link 1

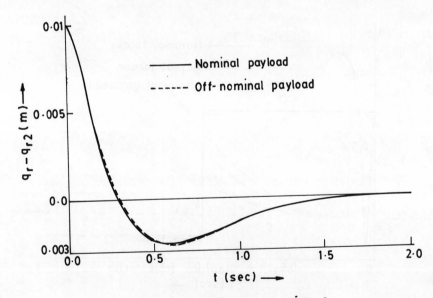

Fig. 3 Tracking error : link 2

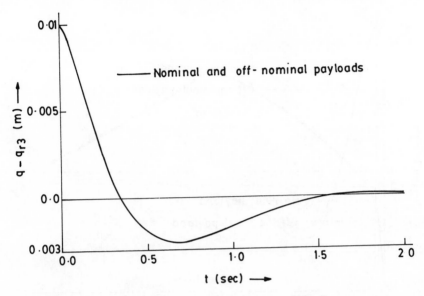

Fig. 4   Tracking error : link 3

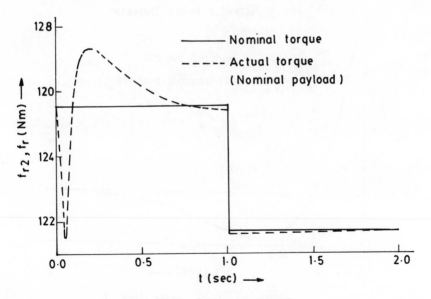

Fig. 5   Torque trajectories

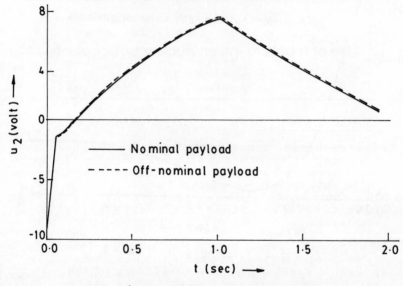

Fig. 6 Actuator input trajectory

Chapter 11

Use of robots in Indian nuclear power plants

S. N. Ahmad and K. Natarajan

1. INTRODUCTION

Pressurized Heavy Water Reactors (PHWRs) have been adopted for the first phase of the Indian nuclear power programme. A distinctive feature of PHWRs is that they have an on - power fuel handling system that permits these reactors to be fuelled even at full power. Long and costly refuelling outages can thus be avoided, leading to better plant capacity and availability factors and superior economic performance.

The fuel handling system in PHWRs is large and complex and consists of the fuelling machine, fuel transfer, spent fuel and auxiliary systems. The fuelling machines which perform the fuelling opera-tions on-reactor are special purpose robots and perform tasks involving accurate positioning, synchronized movement of several devices and applica-tion of high forces. Typical forces applied are 10,000 Kg for clamping and sealing and the typical positioning accuracies required are $\pm$ 0.5mm in a travel of 5300mm and with a moving load of 10 tonnes under conditions of varying catenary pulls. The machines operate at a pressure of 100 Kg/cm2g in a heavy water environment. Fig. 1 shows the fuelling machine for the Narora Atomic Power Plant (NAPP).

The fuelling machines are mobile. This imposes severe restrictions on the size and weight of the equipment that can be used. The premium on space imposes another serious limitation, that back-up (or triplicated channels as in other critical areas of the reactor) cannot be provided. The machines operate in a high radiation environment and are inaccessible during normal operation.

In addition to the above, critical equipment for NAPP is designed to withstand a seismic disturbance as well as withstand high transient temperatures of in a steam laden environment, during plant upset conditions. All these severely restrict the options available to the designer.

In a typical fuelling sequence the two fuelling machines, one having fresh fuel and the other empty,

position themselves at the desired reactor channel
and make a leak tight joint. The machines then remove
the snout, sealing and shielding plugs and in
synchronism load fresh fuel into the reactor and
receive spent fuel. The machines then reinstall the
plugs and the machine with the spent fuel exchanges
spent fuel with new fuel at the fuel transfer system
port. Both machines are now ready for another channel
operation. Meanwhile the fuel transfer system
transfers the spent fuel to the under water spent
fuel storage. Since irradiated nuclear fuel is highly
radioactive all operations are done remotely and
automatically.

2.   OVERALL CONTROL SCHEME

The control system for the fuel handling system
is large, complex and is organized in a hierarchial
manner. It consists of computer, electronic/elec-
trical, fluid power (using several fluids), process
control and mechanical systems. Fig. 2 shows the
overall control scheme for the fuel handling system.
The computer control system, is a distributed
system consisting of a master-computer and Intel 8085
based micro-computers as slaves. The master computer
acts as the supervisor and performs the system
scheduling & maintenance tasks. Each micro controls
one fuelling machine and fuel transfer system. The
master computer schedules and transfers required
codes, including sequential steps, to the micros.
Sequential steps and all process tasks, including
checking of safety logic, are executed in the micros.
The commands from the micros pass through the
hard-wired manual & safety logic, where the essential
safety logic is rechecked, and if satisfied commands
are issued to the fluid power control system. The
fluid power control system then moves the device in
the correct direction at the required speed and force
to perform the desired operation. The feed back
devices are read by the micros and appropriate
control action initiated. The operation and alarm
meassages are displayed on display units and logged
by the printers.
Fuel handling operations are done sequentially
step by step. The system has been designed such that
the fuelling operations can be stopped at any step
for an indefinite period. This is a basic safety
feature, which permits retrieval to be planned should
some maloperation occur. The system has a fail soft
feature which permits three modes of operation namely
auto (full system available), semi-auto (master not
available) and manual (master and micro not
available). The operator can intervene at any point
in the sequence and operate the machine manually.
Operator commands are routed through the hardwired

manual and safety logic system, to prevent gross errors. Computer independent controls and displays have been provided to permit manual operation. The manual and safety logic system has undergone rigorous environmental and seismic testing and analysis to qualify it for reliable operation.

3.   COMPUTER CONTROL SYSTEM

In the configured system the supervisory functions of the master computer are management of the data on the disk, presentation of status on video display units and decoding of commands from the console, for the whole system. The jobs of the microcomputers is thus simplified and 8 bit micros are used. Any of the micros can be operated independently. Information for control of parameters, codes of sequential operation, status for display are required to be transferred between master and micros frequently. The master computer receives commands from operators console. Besides supervisory functions, the master computer also performs on line checks on flow of programmes. A watchdog timer and programmable interval timer assist in this task.

Each microcomputer monitors the status of all field sensors through its digital input multiplexer and analog input subsystems. After processing these signals and performing sequential control functions, the outputs are presented through digital output multiplexers. Six 16-bit interval timers are connected to each micro for synchronization. The watchdog timer provides a check on the proper functioning of the software.

The feedback signals from the field are mainly from switches, potentiometers, linear variable differential transformers,process transmitters and resistance temperature detectors. The digital signals are conditioned through field isolation relays because of the need for high noise immunity and multiple output contacts.

The control system has been so designed that it can be easily restarted after power failure or abnormal shut-down. Important system parameters are stored in memory relays and no storage and no disk or battery back-up is required for the computer system.

As described earlier, the control strategy for both the master and micros is based on periodic intervals by a timer. For control functions the master and micro communicate parallely through command registers and status registers which are wired through DIMs and DOMs. Data transfers, like sequential step logic data, VDU display information and mailbox variables are performed through serial link.

A real time Process Control Language (PCL) has

been developed for writing fuel handling logic
programmes. The programmes in PCL are structured in
blocks of non-executable sequential steps, process
tasks and subroutines. The source programmes are
stored on disk under control of the Text-Editor. The
PCL compiler runs on master computer and generates
executable codes. All object codes generated by the
PCL compiler are executed only in the micro under the
control of the PCL Interpreter programme.

4.   FLUID POWER CONTROLS

Fluid Power Controls involving several fluids
{oil, heavy water (D20) & light water (H20)} are
widely used in the fuel handling system. Hydraulic
actuators have better power/weight ratios, quick
reversibility and excellent stalling characteristics.
Moreover, unlike electrical actuators, torque developed and speed can be independently controlled.
Hydraulic oil is used wherever possible because of
its superior sealing, lubrication and corrosion
resistance properties. Other fluids like D20/H20 are
used when a device operates in a D20/H20 environment
or where chances of contamination exist.
Typically for the fluid power control system
solenoid valves are used for direction control &
hydraulic motors of the bent axis piston type, or
linear actuators,are used for actuation of the
devices. Force selection is done by switching the
PRV reference port to pre-set relief valves.
Speed control is achieved by passing the flow through
pre-adjusted flow control valves. Pilot operated
check valves are used for accurate positioning and
to prevent overhauling when external forces are
applied. Oil supply is from pressure compensated
variable displacement pumps. Pressure line filters
and stainless steel tubes and fittings are used to
obtain a high degree of cleanliness.
While good quality industrial equipment has been
used for oil and H20 system, the use of D20 has posed
special problems and has required the development and
use of special devices. D20 is not only an expensive
fluid, but the D20 used in the fuel handling system
is drawn from the reactor primary coolant and is
radioactive. All D20 equipment is therefore required
to be class I and is designed for zero external
leaks. Since the reactor coolant is almost pure
D20, and no lubricative additives are permitted
because of radiological reasons, the service conditions of D20 equipment are particularly severe from
the fluid power point of view. This has necessitated
the use of custom built equipment for this system
including specially developed solenoid valves, excess
flow check valves, leak detection system, PRVs,
actuation devices and special fittings.

5.   TYPICAL DEVICE CONTROL SCHEME

The control scheme for each device of the fuelling machine has been chosen after due consideration of operational and safety requirements and cover the full range of control schemes such as open loop control, local closed loop control,including servo control, and closed loop control with the computer as a control element.

The control scheme for Ram B is discussed as a typical example. Ram B operates in synchronism with other rams to operate various plugs and handle fuel. Fig. 3 ahows the rams being operated on the reactor sealing plug. Fig. 4 shows the control scheme for the Ram B device. In auto mode the master computer provides the position required to the micro. The micro moves the ram in the correct direction at the desired speed and force, through proper selection of the oil hydraulic devices. The actual position is continuously fed to the micro through a coarse and fine potentiometer combination and a 12 bit analog to digital convertor. Ram B has a travel of 2515mm and the end effector is required to be positioned with an accuracy of $\pm$ 0.75mm under variable loading conditions. The typical force exerted by the ram is 3000 Kg. The ram operates at a high speed of 50mm/sec. and a low speed of 10mm/sec.

6.   CONCLUSION

Fuel handling systems built to the requirements mentioned above have been in operation in the Rajasthan and Madras Atomic Power Plants and have given satisfactory performance. For the Narora Atomic Power Project, due for criticality in Oct.'1988, the system is in an advanced stage of construction. For NAPP, the specifications have been made more stringent and critical equipment has been designed to also meet demanding seismic and plant upset conditions, as discussed earlier.

The major application of robots in Indian nuclear power plants has been in the area of fuel handling . Remotely manipulated tools have also been used in reactor maintenance. For future reactors, the use of robots is being extended for surveillance inspection & maintenance tasks and other areas to reduce the risk to the work force.

7.   LIST OF ILLUSTRATION

Fig. 1   NAPP Fuelling Machine
Fig. 2   Block Diagram of Fuel Handling System Controls.
Fig. 3   Operation of Rams on Seal Plug
Fig. 4   Positioning Loop in NAPP FHC

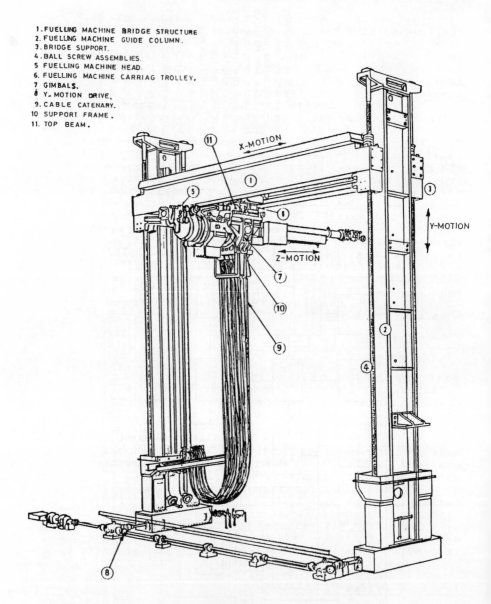

1. FUELLING MACHINE BRIDGE STRUCTURE
2. FUELLING MACHINE GUIDE COLUMN.
3. BRIDGE SUPPORT.
4. BALL SCREW ASSEMBLIES.
5. FUELLING MACHINE HEAD.
6. FUELLING MACHINE CARRIAG TROLLEY.
7. GIMBALS.
8. Y. MOTION DRIVE.
9. CABLE CATENARY.
10. SUPPORT FRAME.
11. TOP BEAM.

FIG·1· NAPP FUELLING MACHINE

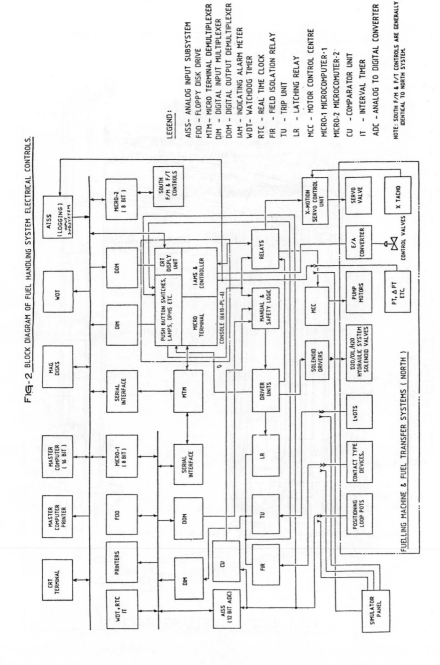

FIG-2 BLOCK DIAGRAM OF FUEL HANDLING SYSTEM ELECTRICAL CONTROLS.

LEGEND:

AISS - ANALOG INPUT SUBSYSTEM
FDD - FLOPPY DISK DRIVE
MTM - MICRO TERMINAL DEMULTIPLEXER
DIM - DIGITAL INPUT MULTIPLEXER
DOM - DIGITAL OUTPUT DEMULTIPLEXER
IAM - INDICATING ALARM METER
WDT - WATCHDOG TIMER
RTC - REAL TIME CLOCK
FIR - FIELD ISOLATION RELAY
TU - TRIP UNIT
LR - LATCHING RELAY
MCC - MOTOR CONTROL CENTRE
MICRO-1 MICROCOMPUTER-1
MICRO-2 MICROCOMUTER-2
CU - COMPARATOR UNIT
IT - INTERVAL TIMER
ADC - ANALOG TO DIGITAL CONVERTER

NOTE: SOUTH F/M & F/T CONTROLS ARE GENERALLY IDENTICAL TO NORTH SYSTEM.

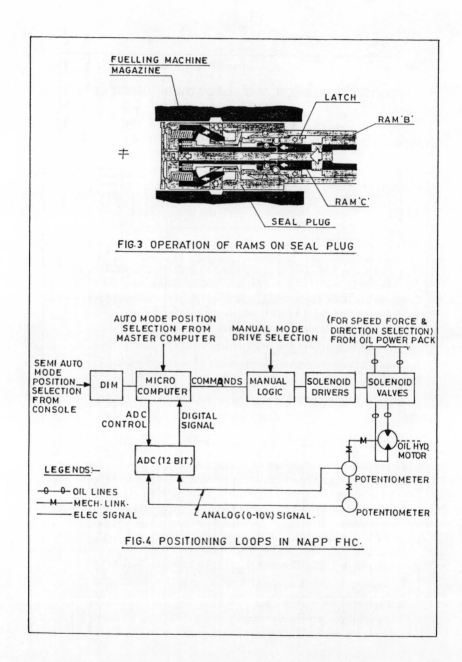

FIG.3 OPERATION OF RAMS ON SEAL PLUG

FIG.4 POSITIONING LOOPS IN NAPP FHC.

Chapter 12

Numerical simulation of a space manipulator control for satellite retrieval

K. Ninomiya, I. Nakatani, J. Kawaguchi, K. Harima,
K. Tsuchiya, M. Inoue and K. Yamada

1. INTRODUCTION

Robotics in space is becoming increasingly important as activities in space expand and sophisticated in-orbit operations are required. We have been proposing an autonomous satellite retrieval experiment using Japanese Space Flyer Unit (SFU).

The SFU project has already been authorized and the first flight is scheduled in 1993. SFU is a re-usable versatile spacecraft which can accommodate various kinds of engineering and scientific experiments. Our experiment is proposed for the 2nd flight of SFU which is expected for mid 1990's.

In this paper, following are presented:

(1) the mission scenario of the proposed experiment,

(2) kinematical and dynamical descriptions of the manipulator system in space,

(3) results of computer simulation and

(4) the method for ground based physical simulation.

2. AUTONOMOUS SATELLITE RETRIEVAL EXPERIMENT

The proposed experiment is autonomous retrieval of a satellite using a manipulator. The target satellite (TS) to be retrieved is assumed to be a "dead" satellite which has stopped all the functions as a satellite. In other words, TS is a passive satellite without orbit and attitude control capability and is assumed to have a tumbling motion.

The SFU has capability of navigation, guidance and control to find, track and rendezvous with the passive TS. The artist conception of the experiment system is shown in Fig.1. The sequence of the experiment is as follows:

(1) SFU is injected into a near earth orbit by H-II rocket.

(2) TS is released from SFU.

(3) After a while, when the distance between SFU and TS is around 10 km, SFU starts searching for TS using a laser radar(Saito et al (1)). TS has corner

cube reflectors on its surfaces and the radar beam scans the space until it finds the reflected beam from TS.

(4) After acquisition of TS by the laser radar, SFU carries out orbit maneuver to rendezvous with TS using the laser radar information. The block diagram of the proposed laser radar is shown in Fig.2.

(5) When the distance of TS and SFU becomes less than 10 m, the laser radar works as a proximity sensor. The corner cube reflectors on the TS surfaces have special patterns so that the tumbling motion of TS can be analyzed by Kalman filter technique using the on-board computer of SFU (Nakatani et al (2)).

(6) Once the identification of the relative position and attitude motion of TS has been complete, SFU takes the best position and attitude for its manipulator to grab the grapple fixture of TS.

(7) The manipulator is autonomously operated and grabs TS using another proximity sensor on the manipulator arm.

(8) At the moment of grabbing TS, the end effector of the manipulator is controlled such that the relative motion of the TS grapple fixture and the end effector is minimum to avoid the excessive reaction force on the manipulator joints.

(9) After completing the acquisition of TS, SFU slowly eliminates the tumbling motion of TS, the manipulator and SFU itself.

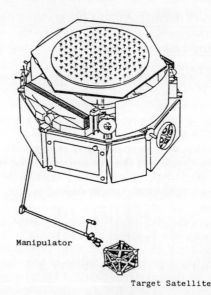

Manipulator

Target Satellite

Fig.1 The concept of retrieval experiment

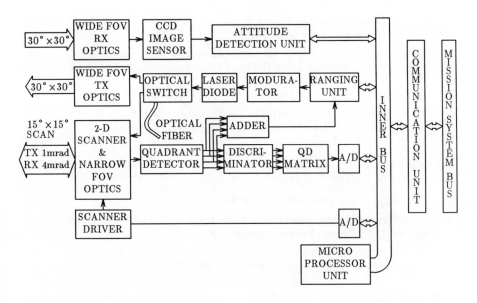

Fig.2 Laser radar block diagram

3. DYNAMICS AND CONTROL OF SPACE MANIPULATOR

3.1 Kinematics and Dynamics

In this chapter, we first explain the kinematics and the dynamics of the manipulator on SFU, comparing with those of manipulators on the ground. Then, we will show an example of a dynamical behavior of SFU by the use of a general simulation program we have developed (Yamada and Tsuchiya (3)).

SFU consists of a satellite main body and a manipulator with n links. We define here the satellite main body as body 0 and links of the manipulator as body $1,2,\cdots,n$ from the shoulder of the manipulator to the hand. The equations of motion of SFU can be derived by using the Kane's equations of motion and are expressed as follows (Yamada and Tsuchiya (3)):

$$\begin{bmatrix} A & 0 & 0 \\ 0 & B & C^T \\ 0 & C & D \end{bmatrix} \begin{bmatrix} \ddot{R} \\ \dot{\omega} \\ \ddot{\Theta} \end{bmatrix} + \begin{bmatrix} X \\ Y \\ Z \end{bmatrix} = 0 \qquad (1)$$

where R is a 3×1 matrix denoting the position of the center of mass of SFU, ω is a 3×1 matrix denoting the angular velocity of the SFU main body, Θ is an n×1 matrix the i-th component of which is the relative rotational angle of the joint between body $i-1$ and body i, A,B,C,D are component matrices of the mass matrix of SFU, and X,Y,Z are matrices which express control forces and torques of SFU and the inertial forces and torques independent of $\ddot{R}, \dot{\omega}, \ddot{\Theta}$.

On the other hand, the momentum P and the angular momentum L of SFU about its center of mass become as follows:

$$P = A\dot{R} \quad L = B\omega + C^T\dot{\Theta} \tag{2}$$

The hand velocity v_h and angular velocity ω_h are expressed as:

$$\begin{bmatrix} v_h \\ \omega_h \end{bmatrix} = G\dot{R} + H\omega + J\dot{\Theta} \tag{3}$$

where G, H, J are coefficient matrices. Now, we will consider the case where the momentum P and the angular momentum L are conserved with the value of 0 during the manipulator operation. Then, the variables $\dot{R}$ and $\dot{\omega}$ can be eliminated from Eq.(1) by the use of Eq.(2) as:

$$(D - CB^{-1}C^T)\ddot{\Theta} + Z^* = D^*\ddot{\Theta} + Z^* = 0 \tag{4}$$

where Z in Eq.(1) is transformed into Z^* by the above elimination. The velocity of hand can also be expressed as (Umetani and Yoshida (4)):

$$\begin{bmatrix} v_h \\ \omega_h \end{bmatrix} = (J - HB^{-1}C^T)\dot{\Theta} = J^*\dot{\Theta} \tag{5}$$

The matrix D^* in Eq.(4) and the matrix J^* in Eq.(5) correspond to the mass matrix and the Jacobian matrix of the manipulator on SFU, respectively, while the matrices D and J correspond to those matrices of the manipulator on the ground. Thus, Eqs.(4) and (5) express the differences of the mass matrix and the Jacobian matrix between these two types of manipulators.

3.2 Control

From Eqs.(4) and (5), the control law of the manipulator on the ground may be applicable to the manipulator on SFU, by replacing the matrices D and J with D^* and J^*. As an example, we show here a simulation result of the resolved acceleration control (Luh et al (5)) of the manipulator on the ground. That is, the joint control torques T of the manipulator are given by

$$T = D^*J^{*\#}(a_d + k_v e_v + k_p e_p) \tag{6}$$

where a_d is the desired acceleration of the hand, e_v and e_p are rate and position errors between the desired trajectory and the hand real motion, k_v and k_p are control gains, $J^{*\#}$ is the inverse matrix of J^* (pseudo inverse is used if J^* is not square.) and the square terms of velocities are omitted. The desired hand trajectory is given so that the hand acceleration a_d is changed linearly with time in the inertial space. Thus, the trajectory of hand position in the inertial space becomes a cubic function of time.

The simulation model of SFU is a satellite with a manipulator of 7 degrees of freedom as shown in Fig.3 and the main parameters of the model are summarized in Table 1. Figure 4 shows an example of the simulation results. The motion of SFU is plotted every second, and the position and orientation of the manipulator hand finally coincide with those of TS grapple fixture.

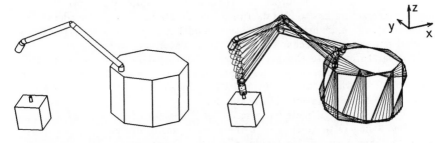

Fig.3 Simulation model   Fig.4 Simulation result

TABLE 1 <u>Main parameters of the simulation model</u>

	SFU	Manipulator						
Body No.	0	1	2	3	4	5	6	7
Length [m]	3×3×2	0.2	2.4	0.2	2.4	0.2	0.2	0.2
Mass [kg]	1000	5	60	5	60	5	5	5

This control law, however, consumes much computation time, and we also applied simpler law by using D and J instead of D^* and J^* in Eq.(6). The result is shown in Fig.5, where the error between the trajectory and the actual hand position in the y-direction is indicated for the case of D and J (solid line) and D^* and J^* (dashed line). The result shows that the simple control law is also effective for the manipulator on SFU. From now on, we will further refine the control law by using the general simulation program and the ground simulation equipment introduced in the next chapter.

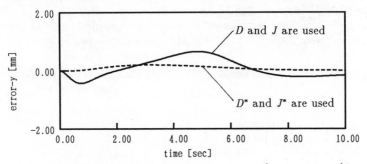

Fig.5 Comparison of control schemes ($k_p = 20$ [sec^{-2}], $k_v = 5$ [sec^{-1}])

4. GROUND SIMULATOR

SFU will retrieve TS automatically using a manipulator which is controlled to follow the path generated by the visual system and the control computer. One of the technical problems is that the motion of the manipulator causes movement in position and attitude of SFU upon which the manipulator and the visual sensors are mounted. In order to control the manipulator correctly, the kinematic interaction has to be taken into consideration. On the other hand, it is difficult to derive a realistic mathematical model including the uncertainties in the visual sensors, joint mechanism of the manipulator, and so on.

In order to evaluate the performances of the system quantitatively, we've designed and constructed a ground simulator. Figure 6 is the outlook of the ground simulator which has 3 translational and 3 rotational degrees of freedom to simulate the relative behavior of SFU and TS.

The manipulator, depicted in Fig.7, has 7 degrees of freedom as a human arm. It has 3 fingers to grasp TS. In the wrist it has a force/torque sensor used to allow the hand a virtual compliance alleviating the shock of collision.

The visual system is composed of 2 video cameras and a signal processor. Figure 8 is the block diagram. The motion of TS is estimated from the 2 dimensional patterns of LED which are allocated on TS.

The specification of the simulator is shown in TABLE 2.

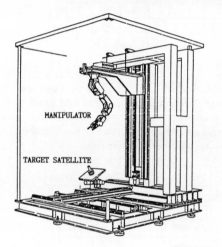

Fig.6 6-axis simulator system

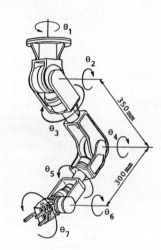

Fig.7 The manipulator with
7 degrees of freedom

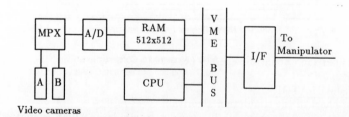

Fig.8 Block diagram of the image processor

TABLE 2 Specification of the simulator

Size	2.8m(width) × 2.9m(depth) × 3.5m(height)	
Weight	3700kg	
Range	X: ±1190mm, Y: ± 800mm, Z: ±1020mm,	θ_x : ±30° θ_y : ±30° θ_z : continuous
Maximum rate	translational: rotational:	300mm/sec 45°/sec
Accuracy	translational: rotational:	0.02mm ±0.01°
Manipulator	degrees of freedom	7
	weight	12.5kg
	length(shoulder to hand)	750mm
	maximum torque	shoulder: 50Nm wrist: 5Nm

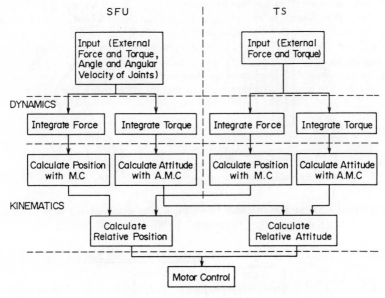

M.C : Momentum Conservation
A.M.C : Angular Momentum Conservation

Fig.9 Flow chart of the simulation algorithm

The relative motion between SFU and TS is simulated based on the law of momentum conservation. The flow of the algorithm is shown in Fig.9. During the period when SFU with the manipulator moves freely without contact to TS, the motions of SFU and the manipulator are simulated as follows: The joint angles and the rates are measured to provide the variation of the momentum and the angular momentum of the manipulator. SFU is to move so that the variation of the momentum and the angular momentum of SFU may have the same values with inverse sign as those of the manipulator. When the manipulator and TS contact with each other, they exchange the momentum and the angular momentum. The amounts of the momentum and the angular momentum may be estimated by the use of a force/torque sensor mounted on TS. Then, the motion of SFU and TS are simulated based on the conservation of momentum and the angular momentum, separately.

5. CONCLUSION

A mission scenario of a satellite retrieval experiment using a manipulator on Space Flyer Unit has been presented. One of the significant features of the space manipulator is that the base is not fixed, but makes a translational and rotational motion during manipulator operations. Manipulator control scheme for the experiment has been given for a movable base in space.

Numerical simulation has been conducted to confirm the validity of the control method. It is found, in the present stage, that the resolved acceleration control works well even when the mass and Jacobian matrices for the manipulator on the ground are used. We are now in the process of refining the control law for the practical use.

A ground based 6-degrees-of-freedom simulator technique has also been presented. We are now conducting a physical simulation using this test facility.

REFERENCES

1. Saito, H., Nakatani, I., Ninomiya, K., and Furuya, A., 1987,
    Proc. 38th Int. Astronautical Federation , 87-53

2. Nakatani, I., Tanamachi, T., and Ninomiya, K., 1986,
    Proc. 37th Int. Astronautical Federation , 86-06

3. Yamada, K., and Tsuchiya, K., 1987, JSME International Journal , 30 , 1667-1674.

4. Umetani, Y., and Yoshida, K., 1986, Proc. 37th Int. Astronautical Federation

5. Luh, J.Y.S., Walker, M.W., and Paul, R.P.C., 1980,
    IEEE Trans. on Automatic Control , 25 , 468-474.

Chapter 13

Computer simulation of a six-legged walking robot using GRASP

R. Naghdy, A. Nott and A. Collie

1. INTRODUCTION

The computer simulation of a robot manipulator has been a popular research and development area for various reasons over the last decade. Off-line programming of a robot for the purpose of accelerating the development of robot application has been perhaps the main reason (Dillman et al (1), Meyer (2), Heginbotham et al (3), Dooner (4)). The facilities provided by the simulation package help the user to select a robot arm for a particular application. The operation of the robot in a specific work-cell in terms of reach, geometry and cycle time may be studied and robot program or its configuration can be modified to produce the desired performance.

The design and development of a robot can also be accelerated by robot simulation. In this case the robot kinematic and dynamic behaviour is studied by the engineer before it is built (Donber et al (5), Leu, Mahjan (6), Stepaneko, Sankar (7), Takanu (8)).

Robot simulation may also replace a real robot in an educational environment (Naghdy (9)). The hands-on experience gained by the students is sufficient to satisfy the educational requirements while the capital cost is relatively low and the safety of the students is not at risk.

Nearly all the robot simulation studies reported in the literature have been based on conventional robot arms bolted down to the factory floor. This work is an attempt to simulate a walking robot platform using a standard simulation package.

Simulation of a walking robot is an interesting and challenging exercise on its own. This perhaps has been the main stimulus behind this work. Similar to the stationary robots, such simulation will aid the designer to apply various control strategies to the robot platform and identify its kinematic and dynamic behaviour.

The work is carried out on GRASP-Graphical Robot Applications Simulation Package. The six-legged robot chosen for the simulation is currently under development as a separate research project at Portsmouth Polytechnic, Robotics Group. The work of solid modelling of the robot is complete. The legs of the robot can move synchronously using the facilities provided in GRASP. More work is still

required to move the platform in conjunction with the
movement of the legs.

In this paper the six-legged robot platform and GRASP
will be initially introduced. The hierarchical structure
adopted for the simulation of the robot will be explained.
The results and the progress up-to-date will be given in the
final section of the paper.

2.   SIX-LEGGED WALKING ROBOT

A mobile robot is a free roving collection of functions
primarily designed to reach a target location in space
(Harmon (9)). The mobile robot developed so far are either
wheeled vehicles or walking machines.

A six-legged walking robot has been under development
for the last two years in the Robotics Group of the
department (Collie et al (10)). The six legs of the robot
are mounted on a roughly coffin shaped body about a meter
long and standing just over half a meter high. The joint
limbs of the robot are powered with compressed air and it
employs compliance and adaptation to achieve smooth fluid
motion over uneven ground.

In the thigh of each leg two single acting cylinders
connect via a steel tape to a drum attached to the lower
leg. The two cylinders are driven in opposition so that a
differential pressure between them will apply a force to
the lower leg. The upper leg is operated by a double acting
cylinder connecting rod and crank. The robot is shown in
Fig.2.1.

Fig.2.1   Six-legged
          walking
          robot

3.   GRASP, A ROBOT SIMULATION PACKAGE

GRASP is an interactive graphical package which models robots and their work place and simulate their operation. It is a development of SAMMIE CAD- a geometric modeller, augmented by facilities such as data base of kinematics models for different robot configurations and robot operating systems (4).

GRASP provides excellent facilities for computer simulation of a robot arm and its work cell. Each object in the work place can be programmed to move and the motions can be synchronised by using variables and flags. In addition a robot is defined as a single unit which can be instructed to move in joint mode, world mode or drive the tool tip onto an object.

The package includes a 3D solid modeller for modelling the flesh of a robot and its work cell. The modeller can be used either interactively or through a descriptive high level programming language. The kinematic limits of the robot and the hierarchical structure of the work cell is also defined through this language.

The entity 'robot' defined in GRASP is a chained linked structure stationary on one end (the base of the robot arm) and open on the other (the tool tip). It will be shown later that such assumption will create problems when simulating a walking robot.

4.   GEOMETRICAL MODEL OF THE ROBOT

The first stage in the simulation of the six legged robot was the elaborate task of geometrical modelling of its structure. The complex body of the robot is basically an assembly of six legs and one platform. The two middle legs and four side legs are identical although the mounting configuration to the platform varies. Due to such similarity it was sufficient to design only two types of legs corresponding to middle and side legs and then multiply them for the rest.

There are two moving links in each leg referred to as lower-leg and upper-leg. Individual links of a leg were modelled geometrically by breaking them down into primitives such as poly prisms, cubes, revsolids (solid of revolution) and cylinders. Primitive objects were defined according to the global frame. They were then shifted and rotated to the correct location during assembly. The assembly of the objects at each stage creates a new object which has the attributes of all its subsets. The model tree of Fig.4.1 illustrates the hierarchy structure of the data generated for upper-leg. The description in each bracket shows the type of the primitive used to model that object. The following instructions show how lower-leg is assembled.

```
SET LOWERLEG = BOTLEG(SHIFT X -47 Y -20.63 Z 34.95)
TOP(SHIFT Y -4.63 Z 34.95 ROTATE X 90)
BOT(SHIFT Z - 265. Y -4.62 ROTATE X 90)
CONN_3(SHIFT Y -2 Z 34.95 ROTATE X 90)
```

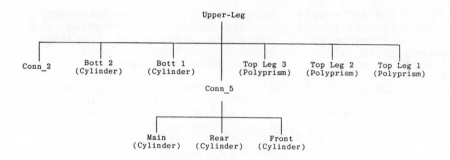

Fig.4.1   Model tree of the upper-leg of the
          six-legged robot

The lower-leg and upper-leg are attached to a mounting panel through a connector. The complete model of one leg is shown in Fig.4.2.

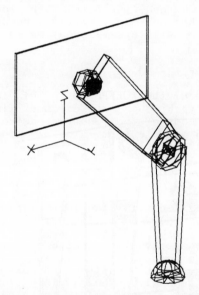

Fig.4.2   The model of one leg of the six-legged robot

In the next stage of the work the body of the robot was modelled and then assembled with the legs.

The data structure in GRASP is arranged in a hierarchical manner. In addition to the geometrical details of the objects and their locations, the data defines the structural relationship in the WORKPLACE. The latter type of data specifies the parent-child connection between every

two entities and shows how the WORKPLACE is organised.   If
object A owns B, the motion of A would move B along.
    The model tree of the WORKPLACE owning the six-legged
robot is shown in Fig.4.3.   Station is the supporting
platform for the robot.   Direction is an arrow indicating
the direction of the movement of the robot.
    The complete model of the robot is illustrated in
Fig.4.4.

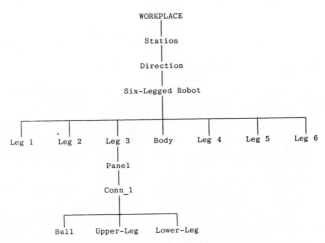

Fig.4.3   Model tree of the WORKPLACE

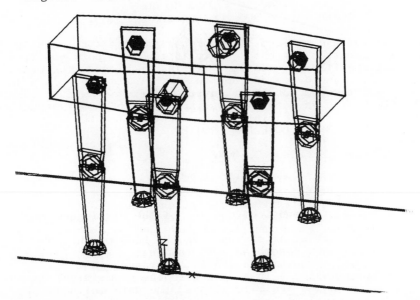

Fig.4.4   The model of six-legged robot on Grasp

5.   THE KINEMATIC MODEL OF THE ROBOT

Functionally each leg of the walking robot consists of two revolute joints at thigh and hip, albeit they are driven by pneumatic cylinders. There is also a third revolute joint in the form of a ball point locating between the upper-leg and the connector to provide side movements for the leg. Since such structure is very similar to a robot arm it was decided to define each leg as a chained linked robot arm (leg) and model it accordingly. Obviously the configuration of this robot is opposite to conventional arms as the tool tip is located at the bottom of the robot and stationary link is at the top.

In GRASP the joint structure of a robot and its kinematic constraints can be easily defined. As an example the robot representing one type of leg is modelled as follows.

```
ROBOT LEG NEW TYPE SIX_LEGGED
JOINT 1 (SHIFT Y 50 Z 26) REVOLUTE X
JOINT 2 (SHIFT Y 38.5 Z 5) REVOLUTE Y
JOINT 3 (SHIFT Z -295) REVOLUTE Y
JOINT 4 (SHIFT Z -325) REVOLUTE X
JOINT 5 REVOLUTE Y
JOINT 6 REVOLUTE Z
TAP
MINIMUM 0 -90 -90 -1 -1 -1
MAXIMUM 7 90 90 1 1 1
INITIAL 0 0 0 0 0 0
PART 7 0 0 0 0 0
VELOCITY 120 120 120 60 60 60
ACCELERATION 200 200 200 200 200 200;
```

The maximum and minimum of joint angles, the top velocity and acceleration for each joint, initial and park positions are all specified by the last six instructions. The dummy joints 4, 5 and 6 are defined to convert the leg to a six-degree-of freedom robot for inverse-kinematic calculations. The flesh of each joint is assigned to it by ADD instruction. LEG_J3 is the first joint of the robot called LEG.

To LEG_J3 ADD LOWERLEG (SHIFT Z - 22.5)

6.   GETTING THE ROBOT TO MOVE

The control system developed for the real six-legged robot is a hybrid force position control. The joint angles are measured by potentiometers attached to the actuators. Silicon sensors monitor the pressures resulting from the pneumatic forces and the ground reaction. An analogue inner control loop maintains the pressure required on the leg while the control computer calculates the demand joint angle and pressure. Obviously the force information is not available for the simulation of the robot on GRASP. Hence the control strategy should rely mainly on the positional information of the joints.

In GRASP all the entities in the WORKPLACE can be programmed to move using the TRACK facilities. A separate track can be specified for each object and the tracks can be run simultaneously. The operation of the tracks is synchronised either by the clock or by employing the variables and flags. The track program can be entered interactively by defining the next location of the objects at each step. Alternatively the high-level text programming language can be used to develop the tracks off-line.

As the first step to simulate the robot it was decided to move the legs in a strategy similar to the control of the real robot. The legs were divided into two trios. The middle leg on one side and the two far legs on the opposite side were grouped as one trio. Each trio supports the robot platform like a tripod while the other three legs move towards the target locations.

According to the control strategy the first trio of the legs, TR1, are lifted off the ground and rotate forwards to pre-determined locations while the other three, TR2, rotate about the hip in the opposite direction to lower the height of the platform for TR1 legs to make contact with the ground. The TR2 legs are now lifted off the ground whilst the TR1 legs rotate backwards about the hip to let TR2 legs make contact with the ground. The sequence is then repeated.

A separate track was written for each leg and the simultaneous operation of the tracks was synchronised by the clock. The first track was chosen as the main track to run in the foreground. The simulation of the movements of the legs although slow and lengthily, was overall successful.

Animating the locomotion of the robot body was the next stage of the work. The movement of the legs should result legitimately in the translation of the robot body in the WORKPLACE. Unfortunately GRASP does not provide any straightforward approach to achieve this aim. In GRASP it is assumed that the base of a robot does not move which is logical for an industrial robot. In six-legged robot however every leg, defined as a chained linked robot, should move entirely and pull the platform along. The attempts so far to find a method have not been successful and more work is required.

7. CONCLUSIONS

The modelling and simulation of a six-legged walking robot on GRASP (Graphical Robot Applications Simulation Package) was reported in this work. The modelling facilities provided in GRASP was sufficient to model the structure and kinematics of the robot successfully. The simulation of the movements of the legs under the robot control strategy was accomplished. The last stage of the work, the locomotion of the robot in the WORKPLACE, requires more work to be fulfilled. GRASP has not been developed for the simulation of the walking robots. Hence some investigation is needed to find a way around this problem.

The dynamics of a robot is not modelled and considered in GRASP. The simulation of the six-legged robot therefore does not fully illustrate the operation of the real machine. The force feedback information from the cylinders is not also available for the simulation unless some effective method is devised to represent this information.

Due to the complexity of the model and amount of the calculations, the simulation of the legs takes considerably a long time. The overall performance of the package gives the impression that the work is stretching GRASP.

REFERENCES

1. Dillmann, R., Hornung, B., and Huck, M., 1986, 'Interactive programming of robots using textual programming and simulation techniques', Proc. 16th Int. Symp. on Industrial Robots, October 1986, Brussels.

2. Meyer, J., 1981, 'An emulation system for programmable sensory robots', IBM Journal of Research and Development Vol. 25, No. 6, November 1981.

3. Heginbotham, W.B., Dooner, M. and Case, K., 1979, 'Rapid assessment of industrial robots' performance by interactive computer graphics', Proc. 9th Int. Symp. on Industrial Robots, Washington DC.

4. Dooner, M., 'Computer simulation to aid robot selection' Robotics Technology Edited by A. Pugh, IEE Control Engineering Series 23.

5. Donbre, E., Fouruier, A., Quaro, C., and Thevenon J.D., 1986, 'Design of a CAD/CAM system for robotics on a microcomputer', Proc. Third Int. Symp. on Robotics Research, MIT Press.

6. Leu, M.C., and Mahajan, R., 1983, 'Simulation of robot kinematics using interactive computer graphics, Proc. ASEE Annual Conference.

7. Stepaneko, Y., and Sankar, T.S., 1985, 'A system approach to dynamic simulation of robotics manipulators' Computer in Mechanical Engineering, May 1985.

8. Naghdy, F., 1987, 'Evaluation of Grasp for teaching robotics to undergraduate students', Proc. CADCAM 87, Birmingham.

9. Harmon, S.Y., 1987, 'Mobile Robots: A tutorial', Tutorial on Mobile Robots, IEEE International Conf. on Robotics and Automation.

10. Collie, A., Billingsley, J., and Hatley, L., 1986, 'The development of a pneumatically powered walking robot base', UK Robotics Research.

Chapter 14

Contouring with pneumatic servo-driven industrial robots

J. Pu and R. H. Weston

1   INTRODUCTION

With the widespread use of computer technology in automating manufacturing functions, new initiatives have emerged in designing pneumatic servos [1]. During the 1980's commercial motion controllers for pneumatic drives have become available from a number of sources worldwide [2,3]. So far, however, the application of pneumatic servos has been limited to the point-to-point positioning of machine mechanisms with very limited control of velocity being possible. Since the release of the first generation pneumatic servos, which were evolved through SERC funded research involving Loughborough University of Technology (LUT) and Martonair UK Ltd [4], research activities have continued with the aim of evolving second-generation pneumatic servos to achieve enhanced performance characteristics. The purpose of this paper is to study the feasibility of achieving contouring using pneumatic drives.

2   EXPERIMENTAL ENVIRONMENT

The experimental system designed and constructed to evolve the control methods is illustrated by Figure 1. The kernal of the control system is the MPE 9995 personal computer chosen in 1984 as it provided adequate software development facilities, fast execution of code, the ability to integrate code derived from both high level and low-level language sources (viz Q-BASIC and TEXAS 9995 assembly language) and offered a stand-alone prototyping machine. Communication between this microprocessor based controller and the interface electronics of the pneumatic drive is achieved through a parallel bus (E-bus) which was introduced by the Texas Instruments Corporation around 1980.

The command signal is converted by the digital to analogue converter (DAC) into an analogue one, which is then amplified by the servo amplifier so that the output signal takes the form of a solenoid current. The current thus produced is supplied to the solenoid which in turn generates a magnetic force at one end of the valve spool to offset the spring. Thus the mass flowrate across the valve can be

controlled by manipulating the spool offset, i.e. through
controlling the current supplied to the solenoid. The load
is moved as the result of a pressure difference between the
two chambers of the actuator. Position and/or velocity
information relating to the piston/load is sensed by a
transducer (rotary or linear) and fed back to the
controller. The pulses produced are encoded and interfaced
to the microprocessor via the position and/or velocity
encoder interface cards, which are mounted on a standard
E-bus backplane. A link between the MPE 9995 machine and a
PRIME 550 mini-computer has also been established so that
the motion and command data can be transferred to the PRIME
where off-line data storage, graphic display and hard-copy
documentation can be carried out. The experimental
conditions under which the test results were obtained and
presented in this paper are tabulated in Table 1.

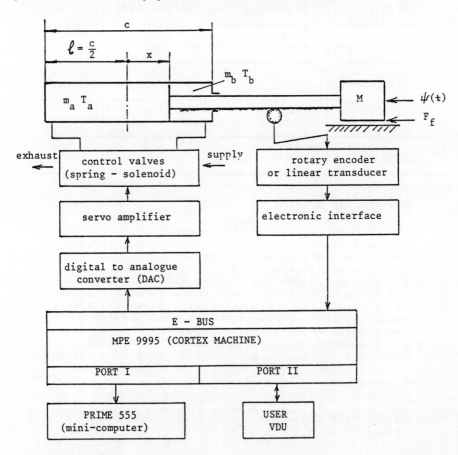

Figure 1   Experimental System

TABLE 1   Experimental Conditions

actuator	Linear cylinder (asymmetric); Stroke 400 mm Bore 25 mm diameter; (Piston rod diameter) 10 mm
servo valve	Single stage; 5 port; solenoid activated
slideway tape	Horizontal
DAC	12 bit (00000 to 01000 hexdecimal)
supply pressure	60 to 100 $\{bf/in^2\}$ or 4 to 7 bar
exhaust pressure	open to the atmosphere
external force	zero
load (kg)	actuator + slideway $\approx$ 25 $\{kg\}$ approximately
friction force	variation: 30 to 50 $\{Newton\}$

3   MODELLING STUDIES

The compressibility of air introduces significant non-linearities into any low pressure pneumatic drive system, which will make such drive systems very difficult to model when compared with hydraulic and electric counterparts [1,5]. Thus, our modelling studies have only been carried out as a means of evolving broad control methods which could be used to accomplish countouring with pneumatic servos. One such modelling study has constructed a second order non-linear model as described by equation (1). In deriving this model viscous friction effects are considered to be negligible and adiabatic flow of an ideal gas is assumed.

MODEL:

$$M\ddot{x} + F_c \, \text{sgn}(\dot{x}) + \frac{R(m_a T_a + m_b T_b)}{\ell^2 (1 - \frac{x^2}{\ell^2})} x = \frac{R(m_a T_a - m_b T_b)}{\ell (1 - \frac{x^2}{\ell^2})} + \psi(t) \quad (1)$$

MASS FLOWRATE [6, 7]:

$$\dot{m} = \dot{m}_c \, f\left(\frac{P_2}{P_1}\right)$$

where $\dot{m}_c = \dfrac{C_d A_o P_1}{\sqrt{T_1}} \sqrt{\dfrac{K}{R}\left(\dfrac{2}{K+1}\right)^{\frac{K+1}{K-1}}}$ (2)

and $f\left(\dfrac{P_2}{P_1}\right) =$

$$\begin{cases} \sqrt{\dfrac{2}{K-1}\left[\left(\dfrac{P_2}{P_1}\right)^{\frac{2}{K}} - \left(\dfrac{P_2}{P_1}\right)^{\frac{K+1}{K}}\right]\left(\dfrac{K+1}{2}\right)^{\frac{K+1}{K-1}}} & \text{subsonic flow when: } \dfrac{P_2}{P_1} > 0.528 \\[4mm] 1 & \text{choked flow when: } \dfrac{P_2}{P_1} < 0.528 \end{cases}$$

note: for air, $K = 1.4$, $\left(\dfrac{2}{K+1}\right)^{\frac{K}{K+1}} = 0.528$

NOMENCLATURE:

$\psi(t)$: external force including any forcing function
   x: displacement from the mid-stroke position
   M: mass of piston including load
   A: area of piston
   C_d: discharge coefficient
   T: absolute temperature
   ℓ: half of the whole length of the two control chambers
   m: gas mass in the control chamber
   R: gas constant
   p: pressure
   F_f: frictional forces
   K: ratio of specific heats
   A_o: area of the valve orifice
      ($A_o = x_o \, b$; x_o is the spool displacement of the valve
                    b is the width of the valve orifice)

SUFFIX:

1: upstream
2: downstream
a: the left side of piston area
b: the right side of piston area

A number of conclusions can be drawn based on the study of this model as outlined in the following subsections.

3.1   Natural Stiffness And Frequency Of The Actuator

Here, natural stiffness and frequency can be derived from the linearised counterpart of equation (1). Assuming small perturbations in position and temperature and that the mass flowrates through the entry and exit ports are nulled (i.e. the mass of air in the control chambers is constant) then the equivalent "spring coefficient" of the system will be:

$$K_s = K_a + K_b = \frac{R \ (M_a T_a + m_b T_b)}{\ell^2(1 - \frac{x^2}{\ell^2})} \tag{3}$$

which shows that the natural stiffness of the system is not only a function of the mass and temperature of the working fluid but also of the position of the piston/load. The natural frequency of the system will thus be: $w_n = \sqrt{k_s/M}$ from which we can see that the system will have a higher natural stiffness towards the ends of the cylinder stroke with the lowest value of stiffness occuring in the mid-stroke position as observed by Burrows [10] and illustrated by Figure 2.

3.2   Transient Response of Velocity to a Step Command

A typical transient response of a pneumatic drive system with a step input is shown in Figure 3. A delay will occur before the pressure difference across the piston is built up to overcome friction, leakage, and so on. Subsequently as the terminal velocity is reached an oscillation is observed [3]. In their research the authors conducted a series of open loop step tests to derive the mean terminal velocity of pneumatic drives whereby a "knowledge" of the drive could be derived for use in contouring control schemes. We will return to this approach later.

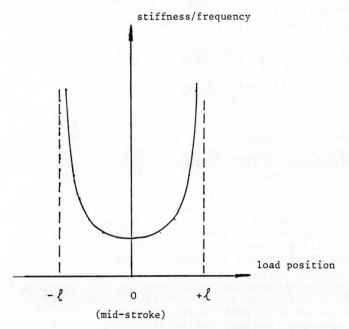

Figure 2 Natural Stiffness/frequency Versus Position
of the Actuator with External Disturbance

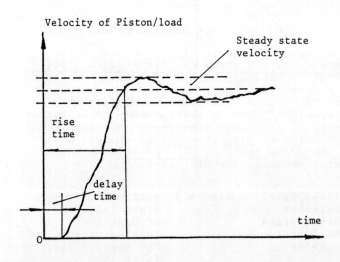

Figure 3   Transient Response to a Step Change in DAC Command

3.3   Possible Valve Configurations

Two alternative control valve schemes have been studied which involve (i) use of two 3-port pneumatic servo valves to independently manipulate the mass flowrates through the entry and exit ports of the actuator or (ii) using one 5-port valve. To achieve a constant speed it is necessary to use two 3-port valves and independently control supply and exhaust flowrates. Only a mean terminal speed can be achieved through using a 5-port valve, as described in 3.2 and illustrated by Figure 3, but a reduction in the cost and complexity of the resulting pneumatic drive and its control software can be achieved. Thus our tests were conducted using one single stage low cost 5-port pneumatic servo valve.

3.4   Step Tests with 5-port Valve

Figure 4 shows the ideal characteristics of a 5-port servo valve. A series of experimental tests were conducted to determine the relationship between the DAC command (to the valve) and the maximum terminal velocity of the piston/load (see Figure 5 from which a deadzone can be observed. It would be instructive if the reader can compare the two curves shown in Figures 4 and 5.

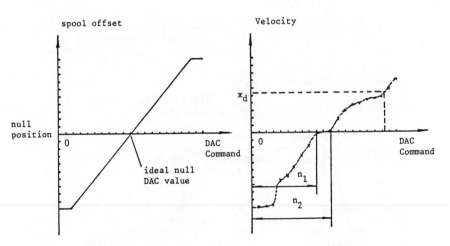

Figure 4   Idealised Spool
             Displacement
             Characteristics
             of the 5-port
             Servo Valve

Figure 5   Maximum Steady
             State Velocity
             of Open-loop
             Testing versus
             DAC Command Step
             Input

4   THE CONTROL LAWS APPLIED

Table 2 compares behavioural characteristics of electric, pneumatic and hydraulic actuators. Using classical small perturbation linearised analysis it can be shown that an electric drive can be represented by a second order model. Correspondingly, pneumatic and hydraulic drives have been represented by 4th and 3rd order linear models respectively. [1, 5, 8].

TABLE 2   Comparison of Behaviour Characteristics between Electric, Pneumatic and Hydraulic Drives

Actuator Type	Null/Condition	Steady State condition for a Step Command	Assumptions
Electric	idle if power switched off	uniform force of torque	frictionless
Pneumatic	oscillation if actuator ports closed	mean velocity with variation	no leakage and friction
Hydraulic	still if actuator ports closed	uniform velocity	no leakage and friction

Present day manipulators (e.g. robots and other manufacturing machines) usually are controlled with very simple control laws which are error driven using some form of equation (4) [8].

$$\text{error command: } d = K_p e + K_v \dot{e} + K_a \ddot{e} + K_i \int e \, dt \tag{4}$$

For electric drives, satisfactory positioning and contouring can be achieved by applying equation (4) in some form to follow velocity profiles of the type shown in Figure 6a [8, 9]. However, in attempting to achieve contouring with pneumatic drives such an approach is not at all appropriate unless high quality pneumatic elements are used (engineered specially to reduce non-linearities) and more complex valving arrangements are devised which would defeat the LUT objective of densing low cost servos which can have widespread industrial application. By referring to Table 2 we see that the pneumatic actuator will tend to oscillate when a single 5-port valve is nulled so that the use of a controlling equation such as equation (4) would fail in following a velocity profile. However, rather than

use a constant null value the DAC command can be made equal
to that value established through open-loop testing i.e. a
valve which corresponds to the mean velocity defined by the
velocity profile being followed (refer to Figures 5 and 7).
Thus when contouring with pneumatic drives a modified
version of equation (4) has been used as described by
equation (5).

error command: $d = k_p e + K_v \dot{e} + k_a \ddot{e} + C$ (5)

where C is obtained from the knowledge based (e.g. Figure
5) and can be referred to as a Null Compensation Value.
Such an approach is critical in accomplishing contouring
when utilising pneumatic drives.

    Choice of k_p, k_v and k_a will not only have significant
effect on the dynamic response and the stability of the
system but will also affect the static performance
characteristics (positioning accuracy) and time lags when
following velocity inputs. Here, decision-making will be
required to determine the weights associated with k_p, k_v
and k_a. The LUT approach here has been to incorporate a
technique of gain scheduling. Figure 8 shows the overall
control strategy evolved.

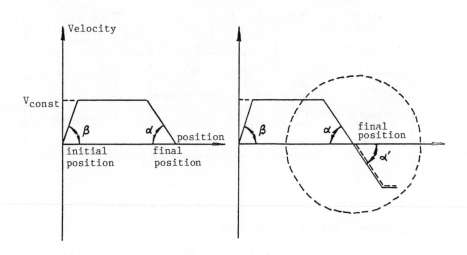

Figure 6a Velocity Profile   Figure 6b Extended Reference
           Defined   Velocity Profile
                                                      for Control

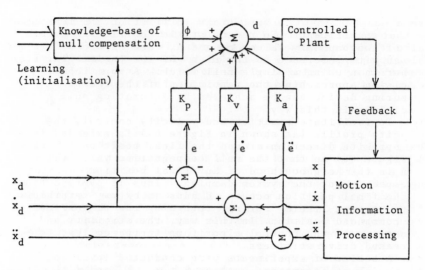

Figure 7   Null Compensation Control Strategy

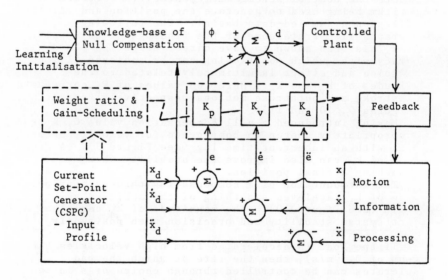

Figure 8   Contouring Control Scheme for Pneumatic Drives

5   ALGORITHMS AND TEST RESULTS

There are various methods by which a "knowledge-base" for the Null Compensation Value C can be established. One practical approach is to build up a experimentally-based look-up table through initialisation and learning procedures involving a limited number of step tests. This approach has proved to be reliable. System hysterisis and

drift, resulting from variations in friction and leakage
can be significant. To reduce static positioning error,
one of two boundary values n_1 and n_2 (see Figure 5) can be
chosen with reference to the sign of the positioning error
rather than using a single null value. A positioning
precision approaching the resolution of the feedback
measuring device can be achieved by automating null
conditions in this way.

To facilitate position and velocity control, the
velocity profile (as shown in Figure 6a) is extended into
the opposite direction around the final position (as shown
in Figure 6b) so that the null compensation value (is used
when in the neighbourhood of any final position
irrespective of the system damping. Thus if overshooting
of the final position occurs, C adds an extra "returning
force" acting like a "spring" returning the piston/load to
its commanded position. In this way, the stability and
robustness of the control algorithms increased lead to
increased drive stiffness.

A series of experiments were conducted involving
the tuning of parameters such as k_p, k_v, k_a and α to
evaluate the control scheme implemented: (a software
algorithm being used to generate the position and velocity
profiles on a sampled-data basis). The following
paragraphs illustrate some of the findings of this work.

k_p has been shown to have direct effect on the
accuracy with which positioning can be achieved [1], but
its choice and effect is ultimately related to the
magnitudes of k_v and k_a. Here two values of k_p correspond
to instroke and outstroke motions respectively (appearing
in Figures 9, 10 and 11).

Having introduced a null compensation value C, K_v can
be appropriately chosen to compensate for velocity errors
albeit with an inherrent time lag (see Figure 9). A proper
choice of k_v can also improve the stability when
approaching a final position.

The introduction of a compensating through appropriate
choice of k_a, achieves smoothing of the achieved velocity
profile (see Figure 10). The magnitude of k_a can also
affect system stability and precision when positioning [1,
3].

If the load is decelerated from high velocities (in
access of 0.5 m/s), then the rate at which the pad
decelerates can be controlled through choice of α in the
generation of command position and velocity profiles. In
such circumstances larger positioning errors can occur
through incorrect choice of the two boundary null values of
the system with the introduction of k_a (a damping factor.
At low velocities the choice of α is less critical.

The above conclusions are based on direct observation
of an experimental system and further evaluation is
required to generalise these results. Furthermore, future
work will be aimed at reducing the inherent time lags
involved when following a velocity profile. Nevertheless,
a level of velocity control has been demonstrated for

pneumatic drives involving the use of readily available low cost components which has not, to the authors knowledge, previously been demonstrated. Here responses for three desired terminal velocities are illustrated in Figure 11. The reader may like to compare the results presented in this paper with those by Rogers and Weston in achieving motion control of electric drives [9]. Requirements in respect to both contouring and positioning being a compromised through the tuning of k_p, k_v and k_a and α.

Velocity (meter/second)

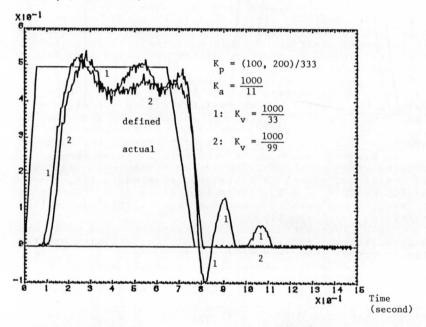

$$K_p = (100, 200)/333$$

$$K_a = \frac{1000}{11}$$

$$1:\ K_v = \frac{1000}{33}$$

$$2:\ K_v = \frac{1000}{99}$$

defined

actual

Time (second)

Figure 9   Effects of k_v on Velocity Profile Output

Velocity (meter/second)

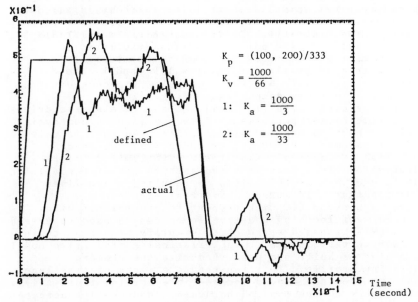

Figure 10 Effects of k_a on Velocity Profile Output

Velocity (meter/second)

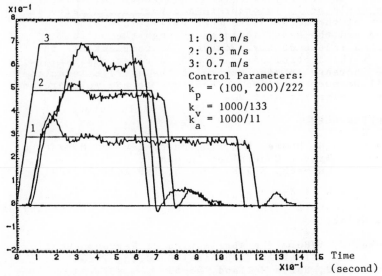

Figure 11 Trapezoidal Velocity Profile Control
(Illustration)

6 CONCLUSIONS AND FUTURE WORK

In low to medium power industrial application areas
(up to say 1 kw), the use of microprocessor-based pneumatic
servos could become popular in the near future as a variety
of new motion controllers become commercially available [3,
4]. Fast and accurate point-to-point positioning can be
achieved with pneumatic modular robots (configured from
single degree of freedom of robot modules) [3]. In
comparison with their hydraulic and electric counterparts,
pneumatic drives can demonstrate an excellent
performance/cost ratio [1, 2, 4]. Although electric drives
can provide better performance in accomplishing contouring
tasks and have become industrially accepted in this role,
low cost pneumatic servo driven industrial robots can
accomplish contouring tasks albeit with limited accuracy
and/or velocities. However, the authors are of the opinion
that future work in this field is required as outlined
below to ensure widespread industrial acceptance of
pneumatic servos: (i) with respect to control software,
enhanced self-learning adaptive loops can be used to extend
this knowledge-based approach to simplify algorithm tuning
and improve performance characteristics; A predictive model
could also be built and used to derive the command
generating profile. This can reduce effects of time lags
resulting from the use of compressible working fluid (ii)
with respect to the control hardware, good quality actuator
and valves can be employed to reduce effects of friction
and leakage (thus reducing system drift and hysterisis) and
to improve the linearity of components (e.g. linearity of
the valve characteristics) so that system behaviour is more
predictable and easily modelled. An example of this
approach would be to incorporate a 3-port servo valve in
each port of the actuator so that time delays can be
minimised and the mass flowrates through the ports
manipulated independently. However, improvements in the
quality of the control hardware should be accomplished with
neglible increase in cost otherwise the average industrial
user will be more inclined to continue to use well proven
higher cost electric drives.

7   REFERENCES

[1]   Weston, R H, Moore, P R, Thatcher, T W and Morgan, C,
       1984, "Computer Controlled Pneumatic Servo Drives",
       Proc. Instn Mech. Engrs, Part B, 198(14), 275-281

[2]   Moore, P R, 1987, "Pneumatic Motion Control System for
       Modular Robots", PhD thesis, Loughborough University of
       Technology

[3]   Backe, W, 1986, "The Application of Servo-pneumatic
       Drives for Flexible Mechanical Handling Techniques",
       Robotics 2, North Holland, 45-56

[4]   Weston, R H and Morgan, G, 1984, "A New Family of robot
      Modules and Their Industrial Application", IMechE Conf.
      on UK Robotics Research, (Mechanical Engineering
      Publication, London)

[5]   McClooy, D and Martin, H R, 1980, "Control of Fluid
      Power: Analysis and Design", 2nd revised edition, Ellic
      Horwood Limited, 337-357

[6]   Sanville, F E, 1971, "A New Method of Specifying the
      Flow Capacity of Pneumatic Fluid Power Valves", Second
      Fluid Power Symposium, BHRA, England

[7]   Shearer, J L, 1954, "Continuous Control of Motion with
      Compressed Air", ScD thesis, Massachusetts Institute of
      Technology

[8]   Craig John J, 1986, "Introduction to Robotics,
      Mechanics, and Control", Wesley

[9]   Rogers, G and Weston, R H, 1987, "A Knowledge-based
      Approached to Robot Axis Control", IEEE International
      Conference on Robotics and Automation, Ralregh, NC, USA

[10] Burrows, G R, 1969, "Effect of Position on the
      Stability of Pneumatic Servo Mechanisms", Research
      notes in Mech Eng Sec., Vol 11, No6, 615-616

Chapter 15

Discrete linear optimal control of robot manipulators

L. G. van Willigenburg

1. INTRODUCTION

The sampling time of a high speed robot motion control system puts a fundamental limit on the control accuracy since the motion cannot be controlled in between two sampling instants. The minimum possible sampling time is determined by the on-line computation time of the control system which depends on the complexity of the control algorithm, which like the on-line computation time determines accuracy of the control. So in the design of robotmotion control schemes one is faced with the problem of trading off controller complexity against on-line computation time.

An articulated arm type robot mechanism is essentially a highly nonlinear system, but the nonlinearities only are significant when the links are moving at high speed. Recently a lot of robot motion control schemes have been presented in literature [12],[13] which deal with the problem of trajectory tracking and treat the robot as a nonlinear system. So in fact these control schemes are designed for high speed trajectory tracking. The sampling time chosen for these robot motion control schemes is hardly ever discussed. The sampling time is only chosen to be small enough not to excite the resonance frequency of the robot mechanism. In practice this means smaller than 20 to 30 ms. If for instance a sampling time of 10 ms is chosen a robot moving at a moderate speed of 1 m/s. travels 1 cm. within 10 ms. Within this 1 cm. there is nothing to control! This very simple example shows that often chosen sampling times around 10 mS seriously limit possibilities for fast and accurate trajectory tracking. The question arises why this phenomena has never drawn serious attention. The reason is that in the design of robot motion control schemes the discrete time character of the computer control, is not taken into account properly. The design results in a continuous time control and the assumption is made that this control may be approximated by a discrete time control.This assumption leads to errors especially in cases of high speed trajectory tracking.

In this paper the problem of trajectory generation and high speed trajectory tracking is treated for Cartesian and Articulated arm type robots. Attention is paid to tracking

of a priori known trajectories and the discrete time
character of the control. Because kinematics and dynamics of
a Cartesian type robot are fare less complicated the on-line
computation time necessary to control this type of robot is
small compared to that necessary to control a Articulated
arm type robot. This offers greater possibilities to reduce
the sampling time which enhances the possible accuracy of
robot motion control.

2.   PROBLEM FORMULATION AND POSSIBLE SOLUTIONS

A robot manipulator consists of a number of connected links
which together we shall call the robot mechanism and in
terms of control the process. Each link is driven by an
actuator, for instance an electrical motor with a gearbox.
The actuators together will be refered to as the robot drive
system. Each actuator is equipped with at least a position
sensor that measures actuator position which is related to
the link position. The actual link position is not measured
since this requires complicated expensive sensors. In case
of perfectly rigid links and drive system the relation
between actuator and link position is straight forward and
simple. Flexibility in links and drive system very much
complicates the relationship which becomes dependent of
forces and torques acting on the process and drive system.
It is often possible to measure actuator speed which is
related to the link speed in a same manner. The speed of an
electrical motor for instance can be measured with a
tachogenerator. This is important since link positions and
speeds together form the state vector of the process. The
robot motion control problem deals with the position control
of each link by applying forces and/or torques to the
process by means of the actuators,given sensor
measurements. The link positions together result in a
position and orientation of a frame attached to the end
point of the final link refered to as the Tool Centre Point
Frame (TCPF). Controlling the TCPF position and orientation
is of primary interest.The dynamics of a N link Articulated
arm robot mechanism can be written as (see [1])

$$\underline{\tau}(t) = M(\underline{\theta})\underline{\ddot{\theta}}(t) + \underline{V}(\underline{\theta},\underline{\dot{\theta}}) + \underline{G}(\underline{\theta}) \tag{1}$$

where

$$\underline{\theta} = (\theta_1,\theta_2,\ldots,\theta_N)^T, \quad \underline{\tau} = (\tau_1,\tau_2,\ldots,\tau_N)^T$$

$\theta_1,\ldots,\theta_N$ are joint angles of the links and $\tau_1,\ldots,\tau_N$
are the actuation torques. $M(\underline{\theta})\underline{\theta}$ represents inertia forces
where M is a NxN mass matrix depending on link positions.
$V(\underline{\theta},\underline{\dot{\theta}})$ is a Nx1 vector dependent on link positions and
speeds representing centrifugal, friction and Coriolis
forces. $G(\underline{\theta})$ is a Nx1 vector depending on link positions
representing gravity forces. (1) is a nonlinear process with
state vector $(\underline{\theta},\underline{\dot{\theta}})^T$. We assume the complete state is
measured and neglect dynamics of the robot drive system. The
LQG control strategy [2,3,4] can be presented in a form that

explicitly deals with the discrete control of a continuous
plant [4,5]. The solution consists of state feedback so it
requires very few on-line computation time which is
important regarding the trade off between on-line
computation time and complexity of the control algorithm.
The LQG design consists of two parts. In the first part an
optimal input and state trajectory, trajectory for short, is
determined which can be done off-line if the desired robot
motion is a priori known (trajectory generation). The second
part deals with the design of a perturbation controller,
based on a linearized perturbation model, that controls the
process as close as possible about this trajectory
(trajectory tracking).

A problem arises if serious process parameter
uncertainty exists. Then one might have to look for adaptive
control strategies. They however require a lot of on-line
computation time and often suffer from stability problems
[6]. Compared to other physical processes the parameters and
dynamics of a rigid robot mechanism are accurately known
[7]. The only parameter uncertainty concerns friction. Many
authors ([7],[12],[13]) also mention parameter uncertainty
concerning the load but in many cases the load is simply
known and otherwise its weight, the main parameter, can be
measured.

A rigid Cartesian type robot mechanism consists of 3
axes moving perpendicular and thus completely independent.
The dynamics can be written as

$$\underline{\tau}(t) = M\underline{\ddot{\theta}}(t) + D\underline{\dot{\theta}}(t) + \underline{G} \tag{2}$$

$$\underline{\theta} = (\theta_1, \theta_2, \theta_3)^T$$

$$\underline{\tau} = (\tau_1, \tau_2, \tau_3)^T$$

(2) is a linear process. M,D are 3x3 diagonal matrices. $M\theta$
represents inertia forces, $D\theta$ forces due to viscous friction
and the 3x1 vector G forces due to gravity. If one of the
axes moves in the direction of gravity two diagonal elements
of G are zero. The state vector of the process is $(\theta, \dot{\theta})^T$.
Since a linear process is just a special case of a nonlinear
process the LQG design is also apliccable to a Cartesian
type robot and results in linear optimal control laws.
The LQG design is presented for both types of robots in
a form that accounts for the discrete time character of the
control.A Cartesian type robot turns out to be much more
attractive when fast and accurate tracking of a priori known
trajectories is required. The main problems in applying the
LQG design to a Articulated arm type robot are
mentioned,which require further research.

3.   LQG ROBOT MOTION CONTROLLER DESIGN

3.1 Partitioning of the control problem

Given a task that the robot has to perform the first

step in the LQG design is the determination of a desired
time evolution of the state,called optimal state trajectory,
corresponding to a time evolution of the input,called
optimal input.The optimal state trajectory and input are a
translation of the task into a specific process behavior and
together are called the optimal trajectory.The determination
of the optimal trajectory is called trajectory generation.
      If there is no process and measurement noise and the
process model is exact,applying the optimal input would
result in exact tracking of the optimal trajectory. Since
for a robot manipulator non of these requirements is
fulfilled it is necesarry to design a controller that will
minimize deviations from the optimal trajectory by means of
feedback. This controller will be refered to as perturbation
controller. Figure 3.1 shows the process-controller
configuration. Note that the optimal input and the
perturbation controller are discrete since the control is
done by a computer. In the LQG design the perturbation
controller is a linear optimal controller which requires few
on-line computations. The design of the perturbation
controller is refered to as the problem of trajectory
tracking.

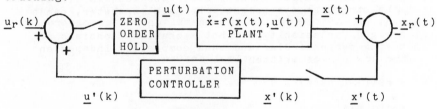

$x_r(t)$:   optimal continuous time trajectory
$u_r(k)$:   optimal discrete time input
u',x' :   perturbation variables

Fig. 3.1 Block diagram of process-controller configuration

3.2   Trajectory generation

      The robot task is usually described by desired positions
and orientations of the TCPF, since the load is carried at
the end point of the final link.Robot kinematics describe
the general relationship between TCPF position and
orientation given positions of the links.

$$\underline{P} = F(\underline{\theta}) \tag{3}$$

For an Articulated arm robot $\underline{P}$ represents TCPF position and
orientation while θ is the vector containing joint angles.
For every θ there is one $\underline{P}$, but for one $\underline{P}$ there can be
several θ. The kinematics are complicated which makes it
impossible to find analytic solutions for θ given an
analytic description of $\underline{P}$.To solve this problem θ is only
calculated for a limited number of TCPF positions and
orientations $\underline{P}_1..\underline{P}_N$.If there is more then one solution,
one solution is chosen.The solutions are connected by an

interpolation strategy (for instance using cubic splines) which results in every joint angle becoming a function of a single parameter λ [8,9].

$$\theta_k = f_k(\lambda) \quad , \quad 0 \leqslant \lambda \leqslant \lambda_{max} \quad , \quad k = 1,2,..,n \tag{4}$$

(4) is a description of the robot task in space coordinates and will be called optimal path. To be able to specify the optimal trajectory the joint angles have to be known as a function of time. This requires the determination of

$$\lambda = g(t) \tag{5}$$

which leads to

$$\theta_k(t) = (f_k \cdot g)(t) \quad , \quad 0 \leqslant t \leqslant t_{max} \quad , \quad k = 1,2,..,N \tag{6}$$

The interpolation strategy which leads to (4) can be the result of an optimization problem constraint for instance by requirements of collision avoidance. If (3) is calculated for many P_k different interpolation strategies will not alter the optimal path significantly in which case the interpolation stategy may be simply chosen. (6) will usually be the result of the following general optimization problem [10] :

Find the continuous time optimal control

$$\underline{u}(t) = \underline{\tau}(t) \tag{7}$$

that minimizes

$$J = K(\underline{x}(t_f),t_f) + {}_0\!\int^{t_f} L(\underline{x}(t),\underline{u}(t))dt \tag{8}$$

where $\underline{x} = (\theta,\dot{\theta})^T$ subject to the dynamics

$$\dot{x} = f(\underline{x}(t),\underline{u}(t)) \tag{9}$$

equality constraints on the state representing the optimal path :

$$\theta_k = h_k.\theta_1 \tag{10}$$

$$\dot{\theta}_k = j_k.\dot{\theta}_1 \tag{11}$$

inequality constraints on the state representing limitations of joint speeds :

$$\underline{V}(\theta) < \underline{\dot{\theta}} < \underline{W}(\theta) \tag{12}$$

and inequality constraints on the input representing bounds on the actuation torques :

$$\underline{S}(\underline{\theta},\underline{\dot{\theta}}) < \underline{\tau} < \underline{Z}(\underline{\theta},\underline{\dot{\theta}}) \tag{13}$$

In literature solutions to this problem have been presented

based on phase plane techniques [8] and dynamic programming [9,10].

The determination of the optimal path and from it the optimal trajectory can be treated as one, very complex optimization problem. This problem is discussed for instance in [11] but no solutions are presented.

The solutions to these optimization problems,if they can be found, all require much computation which however presents no problem if the robot task is a priori known since then the computation can be done off-line. All solutions however assume a continuous time control. Since we have a discrete time control it generally is impossible for the plant to follow the optimal path (4). What is done in practice is that the discrete time optimal input $u_r(k)$ is taken equal to the continuous time optimal input, corresponding to the trajectory determined by (6), at the sampling instants. But this means that the discrete time input is suboptimal. The suboptimality of the discrete time control becomes significant in cases of high speed trajectory tracking.An optimal trajectory corresponding to a discrete time optimal input will be the solution of the general optimization problem (7)..(13) with

$$\underline{u}(t) = \underline{u}(k) \ , \ k.T_s \leqslant t < (k+1).T_s, k = 0,1,..,m \qquad (14)$$

$$t_f = m.T_s \qquad (15)$$

where T_s is the sampling time.The optimization problem deals with the optimal discrete time control of a continuous time process. Since we have a discrete time control the process in general cannot track the optimal path which in this case cannot be represented by equality constraints (10),(11). One possibility is to use the solution of (7)..(13) which asumes a continuous time control. If we call the optimal state trajectory $\underline{x}_0(t)$ then we can state the following optimization problem :

Find the discrete time control (14) which minimizes

$$J = K(\underline{x}(t_f),t\ ) + {}_0\!\int^{t_f} L(\underline{x}(t)-\underline{x}_0(t)),\underline{u}(t))dt \qquad (16)$$

subject to (9),(12),(13) and (15)

The solution of the problem in which a continuous time control is assumed serves now as a reference for the problem in which there is a discrete time control. The differences of the solutions to (7)..(13) and (14),(16) indicate the errors due to approximating the continuous time control by a discrete time control. For an Articulated arm robot the optimization problem (14),(16) is very complex and has to be a subject of further research.

The kinematics of a rigid Cartesian type robot are trivial. The orientation of the TCPF is always the same. We are only concerned with the position of the Tool Centre Point (TCP). If this position is described in a Cartesian frame with the axes parallel to the robot links the

translation θ of each link determines one of the TCP
coordinates. So an analytical description of the TCP
coordinates equals the optimal path. In practice
however,often only a limited number of TCP positions are
known,for instance from an image or task discription. Then
again an interpolation strategy has to be determined to
specify the optimal path.

Assuming a continuous time control, given the optimal
path, the optimal trajectory is the result of solving
(7)..(13). Because the rigid Cartesian robot is a linear
decoupled process the equality and inequality constraints
(10)..(13) are less complicated.

Solving (14),(16) for a linear process is a standard
optimization problem if the costcriterion (16) is quadratic
(see [4],[5])! Given the noise free Equivalent Discrete time
System (EDS) which discribes the process behavior of the
linear continuous time process at the sampling instants :

$$\underline{x}(k+1) = \Phi(k)\underline{x}(k) + \Gamma(k)\underline{u}(k) \tag{17}$$

and the costcriterion

$$J = {}_0\!\int^{tf} \left[\underline{x}(t)^T Q(t)\underline{x}(t) + \underline{u}^T(t)R(t)\underline{u}(t) \right]dt \tag{18}$$

then

$$\underline{u}(k) = F\ \underline{x}(k) \tag{19}$$

minimizes (18) subject to (17) where

$$F(k)=\widetilde{R}^{-1}(k)\widetilde{M}^T(k)+(\widetilde{R}(k)+\Gamma^T(k)\Sigma(k)\Gamma(k))^{-1}\Gamma^T(k)\Sigma(k)\widetilde{S}(k)$$

where $\Sigma(k)$ is the solution to the matrix riccati equation

$$\Sigma(k)=\widetilde{Q}(k)-\widetilde{M}(k)R^{-1}(k)\widetilde{M}^T(k)+\widetilde{S}^T(k)\left[\Sigma(k)-\Sigma(k)\Gamma(k)\widetilde{R}(k)+\right.$$

$$\left.\Gamma^T(k)\Sigma(k)\Gamma(k))^{-1}\Gamma^T(k)\Sigma(k)\right]\widetilde{S}(k)$$

where $\widetilde{R},\widetilde{M},\widetilde{Q},\widetilde{S}$ are matrices corresponding to the equivalent
discrete optimal control problem (see [4],[5])

3.3 Trajectory tracking

The perturbation controller (figure 1) generates an
input $\underline{u}'(k)$ with the general objective to keep both $\underline{u}'(k)$
and $\underline{x}'(\overline{t})$ small. The relationship between $\underline{u}'(k)$ and $\underline{x}'(k)$ is
a discretization of the relationship between $\underline{x}'(t)$ and
$\underline{u}'(t)$. In the case of a nonlinear process, the articulated
arm robot, using a Taylor series expansion about the
trajectory $\underline{x}_r(t)$, $\underline{u}_r(t)$ neglecting the higher order
terms gives the linearized perturbation model (20)
describing aproximately the relationship between $\underline{x}'(t)$ and
$\underline{u}'(t)$.

$$\underline{x}'(t) = A(t)\underline{x}'(t) + B(t)\underline{u}'(t) \tag{20}$$

where A(t) and B(t) are jacobian matrices

$$A(t) = \frac{\partial f}{\partial \underline{x}}\bigg|_{\underline{x}_r(t)} \quad (21) \qquad B(t) = \frac{\partial f}{\partial \underline{u}}\bigg|_{\underline{u}_r(t)} \qquad (22)$$

(22) represents a continuous time,time variable process, which depends on the trajectory $\underline{x}_r(t), \underline{u}_r(t)$. The remaining problem is to control this process about the zero state using the discrete time control $\underline{u}'(k)$. The problem again comes down to (17), (18) which determines the optimal discrete time control of a continuous time linear process. Note that the costcriterion (18) is quadratic and minimizes errors due to linearization. This minimization can be influenced by the design parameters Q(t) and R(t) .

The computation of (21) , (22) poses a serious problem. For a simple two link planar manipulator some elements of A(t) and B(t) contain over 50 components. Although computation of the linearized perturbation model involves a great amount of computation it should be stressed that it is possible to calculate A(t), B(t). The computation can be done off-line so if a computer program could be written to perform the linearization (21), (22) the problem would be solved.

In case of a Cartesian type robot, a linear time invariant process, linearizing the perturbation model about the optimal trajectory is not necesarry. In this case the perturbation model, and so the optimal perturbation control law, do not depend on the optimal trajectory.

4. CONCLUSIONS

The design of fast and accurate robot motion control schemes should include a discription of the discrete time character of the control. The LQG design can be presented in a form that explicitly deals with the discrete time perturbation control of a continuous time process about an off-line determined trajectory. This trajectory should consist of a state trajectory corresponding to a discrete time control. If this trajectory is determined by an optimization problem this problem has to account for the discrete time character of the control.

Solutions to the problem of optimal trajectory generation for robotmotion, asuming a continuous time control, have been presented in literature. These solutions can be used to optimize trajectories corresponding to a discrete time control. For a Cartesian robot, a linear process, this optimization has a known solution if the costcriterion is quadratic. For a Articulated arm type robot this optimization presents a very difficult problem that requires further research.

The optimal perturbation control for a Cartesian robot can be easily determined. The design of the optimal perturbation controller for a Articulated arm type robot involves a linearization procedure which poses a serious problem and should also be a subject of further research.

REFERENCES

1. Craig J.J., 1986
    Introduction to Robotics Dynamics and Control , Addison
    and Wesley
2. Athans M., 1971
    'The Role and Use of the Stochastic Linear Quadratic
    Gaussian Problem in Control System Design',IEEE
    Transactions on Automatic Control Vol AC 16 pp.
    529-553
3. Athans M., 1972
    'The Discrete Time Linear Quadratic Gaussian Stochastic
    Control Problem', Annals of Economic and Social
    Measurement Vol 2, 1/4 pp. 449-491
4. Johnson A., 1985
    Process Dynamics Estimation and Control, Peter
    Peregrinus
5. De Koning W.L., 1980
    'Equivalent discrete optimal control problem for
    randomly sampled digital control systems',
    International Journal of System Science, 11, 7,
    pp. 841,850
6. Ortega R., Tang Y., 1987
    'Theoretical results on robustness of direct adaptive
    controllers : A survey', Preprints IFAC 10th World
    Congres on Automatic Control, Vol. 10, pp. 1-16
7. Vukobratovic M., Kircanski N., 1985
    'An Approach to Adaptive Control of Robotic
    Manipulators'
    Automatica Vol 21, 6, pp. 639-647
8. Shin K.G., Mc Kay N.D., 1985
    'Minimum Time Control of Robotic Manipulators with
    Geometric path constraints', IEEE Transactions on
    Automatic control Vol AC 30, 6,pp. 531-541
9. Shin K.G., Mc Kay N.D., 1986
    'A Dynamic Programming Approach to Trajectory Planning
    of Robotic Manipulators, IEEE Transactions on
    Automatic Control Vol AC-31, 6,pp. 491-500
10. Singh S., Leu M.C., 1987
    'Optimal Trajectory Generation for Robotic Manipulators
    Using Dynamic Programming', Transactions of the ASME
    Journal of Dynamic Systems Measurement and Control
    Vol 109, pp. 88-96
11. Geering H.P.,Guzzella L.,Hepner A.R., Onder C.H., 1986
    'Time Optimal Motions of Robots in Assembly Tasks',
    IEEE Transactions on Automatic Control Vol AC-31, 6,
    pp. 512-518
12. Lee C.S.G., Chung M.J. ,1984
    'An adaptive Control Strategy for Mechanical
    Manipulators', IEEE Transactions on Automatic Control
    Vol AC-29, 9,pp. 837-840
13. Choi Y.K., Chung M.J., Bien Z., 1986
    'An adaptive control scheme for robot manipulators'
    International Journal of Control Vol 44, 4,
    pp.1185-1191

Chapter 16

Achieving singularity robustness: an inverse kinematic solution algorithm for robot control

P. Chiacchio and B. Siciliano

1. PROBLEM STATEMENT AND PREVIOUS WORK

1.1 Robot Kinematics

For any robot with known geometrical dimensions, the direct kinematic equation describes the mapping of the ($n \times 1$) vector of joint coordinates $\underline{q}$ into the ($m \times 1$) vector of robot's end-effector task (Cartesian) coordinates $\underline{x}$ as (Denavit and Hartenberg (1))

$$\underline{x} = \underline{f}(\underline{q}) \qquad (1.1)$$

where $\underline{f}$ is a continuous nonlinear function, whose structure and parameters are known. Differentiating Eq.1.1 with respect to time yields the mapping between the joint velocity vector $\underline{\dot{q}}$ and the end-effector task velocity vector $\underline{\dot{x}}$, through the ($m \times n$) Jacobian matrix $J(\underline{q}) = \partial \underline{f}/\partial \underline{q}$, i.e.

$$\underline{\dot{x}} = J(\underline{q})\underline{\dot{q}}. \qquad (1.2)$$

1.2 The Inverse Kinematic Problem

A robot is usually commanded by assigning a desired motion $\underline{\tilde{x}}(t)$ to its end-effector. Hence, it is necessary to solve Eq.1.1 for $\underline{\tilde{q}}(t)$ such that a control system can be designed which guarantees tracking of the desired joint motion $\underline{\tilde{q}}(t)$. Solving the inverse kinematic problem becomes more dramatic for on-line applications, when the end-effector's motion is re-programmed on the basis of sensor information.

1.2.1 Previous work. The most natural way of solving the inverse kinematic problem relies upon the possibility of finding a closed-form analytical solution to Eq.1.1. Pieper (2) shows that this is true only for nonredundant structures ($m = n$) having simple geometries. In particular, a sufficient condition is given which establishes that the kinematic structure is solvable if it contains three consecutive joint rotation axes intersecting at a common point. For instance, all the robots having spherical wrists are solvable; there do exist, however, mechanical designs that do not satisfy the above condition. In addition, if the joint velocities are needed by the control servos, Eq.1.2 must be solved for $\dot{q}$, thus requiring also the inversion of the Jacobian matrix $J(q)$. Therefore, the two shortcomings of the above technique, namely the solvability of the structure and the computational burden, have inspired the research to finding alternative solution techniques to the inverse kinematic problem which be applicable to any kinematic structure as well as be efficient from the computational viewpoint.

The other approach to the problem, commonly followed in the robotics literature, is based on the use of the inverse of the Jacobian matrix in Eq.1.2. It does not require any special assumption on the kinematic structure. In particular, it can be shown that the general solution to Eq.1.2, for a kinematically redundant structure (m < n), is given by (Whitney (3))

$$\dot{q} = J^{\dagger}\dot{x} + [I - J^{\dagger}(q)J(q)]\dot{q}_0 \qquad (1.3)$$

where $J^{\dagger}$ is the (n x m) Moore-Penrose pseudo-inverse matrix defined as $J^{\dagger} = J^T(JJ^T)^{-1}$, I is the (n x n) identity matrix and $\dot{q}_0$ is an (n x 1) arbitrary joint velocity vector. It can be noticed that the solution given in Eq.1.3 composes of the least-square solution term of minimum norm plus a homogeneous solution term created by the projection operator $(I - J^TJ)$ which selects the components of $\dot{q}_0$ in the null space (space of redundancy) of the mapping J. The vector $\dot{q}_0$ is usually adopted to optimize some additional criterion, such as obstacle avoidance, limited joint range etc. (see Klein (4) for a short survey on various choices of the vector $\dot{q}_0$).

1.2.2 Kinematic Singularities. The other drawback encountered in the solution to the inverse kinematic problem is certainly the occurrence of a kinematic singularity. A joint configuration is detected as a singular configuration if the determinant of the Jacobian matrix in Eq.1.2 vanishes. In case of redundant structures, a joint configuration is singular if the Jacobian matrix is not a full rank matrix. It can be seen that at a singularity either one or more columns of the Jacobian are null vectors or there is colinearity between two or more columns. The former case corresponds to having one or more joints whose motions do not produce any change of the end-effector location, like the "shoulder singularity" in a PUMA-like geometry (Hollerbach (5)). This type of singularity is the most critical one since it may fall into the robot's reachable workspace, thus constituting a problem for end-effector correct motion planning. On the other hand, in the latter case two or more joint infinitesimal motions produce the same infinitesimal change of the end-effector location, like the "elbow singularity" in a PUMA-like geometry (5). This kind of singularity is not as bad as the former since in those particular configurations the end-effector is at the boundaries of the robot's reachable workspace. Another singularity which usually occurs for manipulators having a spherical wrist is known as "wrist singularity" (5, Aboaf and Paul (6)), at which the wrist cannot accomodate rotations about one of its orientation axes, which then becomes a degenerate axis.

The main problem concerned with a solution of the kind of Eq.1.3 is that kinematic singularities are not avoided in any practical sense, since the joint velocities are minimized only instantaneously (Baillieul et al (7)). Nevertheless, a solution based on Eq.1.3 with singularity robustness has been recently proposed by Nakamura and Hanafusa (8).

2. AN INVERSE KINEMATIC SOLUTION ALGORITHM WITH SINGULARITY ROBUSTNESS

The inverse kinematic solution algorithm with singularity robustness, which is the main contribution of this work, is naturally derived from a general computational method recently established in the literature (Balestrino et al (9), Siciliano (10), Balestrino et al (11)) which is applicable to any redundant and nonredundant robot geometry. Hence, the next subsection is devoted to briefly recall that general algorithm.

2.1 The General Inverse Kinematic Solution Algorithm

The inverse kinematic problem is solved by constructing the dynamic system of Fig.2.1, whose input is the end-effector target trajectory $\tilde{x}(t)$ and whose outputs are the corresponding joint position and velocity trajectories, $q(t)$ and $\dot{q}(t)$ respectively; K is a positive definite diagonal matrix.

Lemma. The dynamic system of Fig.2.1 assures that the tracking error $\underline{e}(t) = \tilde{x}(t) - \underline{x}(t)$ can be made arbitrarily small by increasing the minimum element of K.

Proof. According to the Lyapunov direct method for the analysis of the stability of nonlinear systems, define the positive definite Lyapunov function

$$v(\underline{e},t) = \tfrac{1}{2}\underline{e}^T(t)\underline{e}(t). \tag{2.1}$$

Its time derivative results, via Eq.1.2 (dropping the time dependence)

$$\dot{v}(\underline{e}) = \underline{e}^T\dot{\tilde{x}} - \underline{e}^T J(q) K J^T(q)\underline{e}. \tag{2.2}$$

It can be recognized that $\dot{v}(\underline{e})$ is negative definite only outside a region in the error space containing $\underline{e} = \underline{0}$, which is attractive for all trajectories. The maximum tracking errors will depend directly on $\|\dot{\tilde{x}}\|$ and inversely on the minimum element of K. It must be emphasized that the steady-state error ($\dot{\tilde{x}} = \underline{0}$) is identically zero.

From the above lemma it follows that the application of the dynamic system of Fig. 1 to solve the inverse kinematic problem for a general structure is twofold. It can be used off-line to make $q(t)$ approach a desired constant solution $\hat{q}$ to Eq.1.1, with $q(0) \neq \hat{q}$, arbitrarily fast. It can be adopted on-line to guarantee that $\underline{x}(t)$ will track the desired end-effector trajectory $\tilde{x}(t)$ with an arbitrarily fast

decaying error.
     The advantages of this technique can be summarized as:
a)   it is applicable to any robot since it does not require
     any special assumption regarding the kinematic struc-
     ture,
b)   it is computationally efficient since it is based only
     on direct kinematic functions ($\underline{f}$ and J), generating
     joint velocities at no additional cost,
c)   the use of the transpose of the Jacobian may avoid
     problems when kinematic singularities occur (this point
     will be faced up in the following subsection),
d)   given the initial configuration of the structure,
     uniqueness of the solution is assured as the algorithm,
     generates adjacent solutions step-by-step.
The same algorithm can be partitioned into two stages to
better account for the particular geometry of the structure,
with the inherent advantage of further decreasing the computational burden ((9), (10), (11), De Maria et al (12),
Sciavicco and Siciliano (13,14,15).

2.2 Making the Algorithm Robust to Singularities

     Based on the remarks of subsection 1.2.2, only
kinematic singularities that cause one column of the
Jacobian matrix of Eq.1.2 to vanish are considered in what
follows. Thus, assume that $\underline{j}_i$ be the null column vector of
J. This implies that the motion of the corresponding joint
q_i does not produce any change of the end-effector location.
Similarly, if the computational scheme of Fig.2.1 is adopted
to solve the inverse kinematic problem, it is easy to recognize that there will be no motion at the joint q_i. This result is consistent with the mechanical interpretation that
it is not worth moving the joint q_i.
     The real drawback to the solution of the inverse
kinematic problem, however, occurs when the trajectory assigned to the end-effector passes in the proximity of a singularity. In that case, indeed, the norm of the vector $\underline{j}_i$
approaches zero and higher values for the corresponding
joint velocity $\dot{q}_i$ are expected to allow the end-effector to
track the desired trajectory. On the other hand, from the
point of view of the scheme of Fig.2.1 with constant gains
in the matrix K, it happens that the weight of the tracking
error $\underline{e}(t)$ on the control $\dot{q}_i$ is "masked" by the small value
of the norm of $\underline{j}_i$, compared to the other joints. This implies that the joint q_i cannot move as fast as required by
the end-effector trajectory and the tracking error tends to
increase.
     In order to overcome the above problem, here it is pro-
posed to adjust the elements k_i of the matrix K of the
scheme of Fig.2.1 according to the current joint configuration, such that the algorithm be robust to the occurrence of
kinematic singularities. The suggestion is to modify the
elements k_i into

$$k_i / \| \underline{j}_i(\underline{q}) \| \qquad (2.3)$$

which assures that, in the proximity of a singularity when $\|\dot{j}_i\|$ takes a small value, only the weight for the control $\dot{q}_i$ increases, guaranteeing a contained tracking error. Obviously, if the trajectory crosses the singular point, the above weight is not allowed to take an infinitely large value. In other words, there must be a numerical threshold for $\|\dot{j}_i\|$ so as to avoid division by zero in Eq.2.3.

It has to be emphasized that, in light of the choice in Eq.2.3, the control at each joint becomes

$$\dot{q}_i = (k_i/\|j_i(q)\|)j_i^T e \qquad (2.4)$$

which corresponds to making the actual weight quasi-independent (see the angle of the inner product $j_i^T e$) of the particular configuration q attained by the structure along the trajectory.

This point turns out to be advantageous for the discrete-time implementation of the algorithm. It can be recognized, indeed, that there does exist a maximum value for the equivalent gains of K which depends inversely on the sampling time. To this purpose, the above "normalization" serves as a design tool to set the k_i's regardless of the desired end-effector trajectory to track.

3. A PRACTICAL EXAMPLE

In order to show the effectiveness of the proposed inverse kinematic solution algorithm with singularity robustness, a case study has been worked out. The robot prototype DEXTER available at FIAR S.p.A., Italy (Fig.3.1) has been selected. It has seven degrees of freedom (redundant) and a PUMA-like geometry as regards the joints q_2, q_3 and q_4. Only the first four joints are considered in the two sets of simulations that follow. It is quite straightforward to recognize that a shoulder singularity occurs at any point along the axis of the shoulder joint q_2.

In the first set, the sliding joint q_1 is assumed to be blocked such that the structure be nonredundant for an end-effector positioning task. The desired trajectory is a straight line parallel to the floor, 15 cm. away from the above singularity axes. Fig.3.2a shows that the end-effector position tracking error becomes considerably smaller if the modification of the gains k_i's (Eq.2.3) occurs. The shoulder joint q_2 anticipates its motion, according to the increased sensitivity to the tracking error (Fig.3.2b).

In the second set, q_1 is released such that the structure becomes redundant for the same kind of task as above. The desired trajectory is a straight line parallel to the floor, crossing the plane formed by the axes of the sliding joint q_1 and the shoulder joint q_2, with the peculiarity that the trajectory does have a component on the axis of q_1, which is then required to slide. An improvement on the tracking error can be seen (Fig.3.3a) as in the previous case. In addition, the joint velocity $\dot{q}_1$ decreases as all the joints concur to better accomplish the required motion of the robot's end-effector (Fig.3.3b).

REFERENCES

1. Denavit, J., and Hartenberg, R.S., 1955, 'A kinematic notation for lower-pair mechanisms based on matrices', ASME J. Appl. Mech., 22, 215-221.

2. Pieper, D.L., 1969, 'The Kinematics of Manipulators under Computer Control', Ph.D. dissertation, Stanford University, U.S.A..

3. Whitney, D.E., 1972, 'The mathematics of coordinated control of prosthetic arms and manipulators', ASME J. Dyn. Syst., Meas., Contr., 94, 303-309.

4. Klein, C.A., 1985, 'Use of redundancy in the design of robotic systems', Robotics Research: The Second International Symposium, MIT Press, 207-214.

5. Hollerbach, J.M., 1985, 'Evaluation of redundant manipulators derived from the PUMA geometry', Proc. of the ASME Winter Annual Meeting, 187-192.

6. Aboaf, E.W., and Paul, R.P., 1987, 'Living with the singularity of robot wrists', Proc. of the 1987 IEEE International Conference on Robotics and Automation, 1713-1717.

7. Baillieul, J., Hollerbach, J., and Brockett, R.W., 1984, 'Programming and control of kinematically redundant manipulators', Proc. of the 23rd IEEE CDC, 768-774.

8. Nakamura, Y., and Hanafusa, H., 1986, 'Inverse kinematic solutions with singularity robustness for robot manipulator control', ASME J. Dyn. Syst., Meas., Contr., 108, 163-171.

9. Balestrino, A., De Maria G., and Sciavicco L., 1984, 'Robust control of robotic manipulators', Prep. of the 9th IFAC World Congress, 6, 80-85.

10. Siciliano, B., 1986, 'Solution Algorithms to the Inverse Kinematic Problem for Manipulation Robots (in Italian)', Ph.D. dissertation, Università di Napoli, Italy.

11. Balestrino, A., De Maria, G., Sciavicco, L. and Siciliano, B., 1987, 'An algorithmic approach to coordinate transformation for robotic manipulators', Advanced Robotics, 2, to be published.

12. De Maria, G., Sciavicco, L. and Siciliano, B., 1985, 'A general solution algorithm to coordinate transformation for robotic manipulators', Proc. of the Second International Conference on Advanced Robotics, 251-258.

13. Sciavicco, L., and Siciliano, B., 1986, 'An inverse kinematic solution algorithm for robots with two-by-two intersecting axes at the end effector', <u>Proc. of the 1986 IEEE International Conference on Robotics and Automation</u>, 673-678.

14. Sciavicco, L., and Siciliano, B., 1986, 'Coordinate transformation: A solution algorithm for one class of robots', <u>IEEE Trans. Syst., Man, Cybern.</u>, <u>16</u>, 550-559.

15. Sciavicco, L., and Siciliano, B., 1986, 'Solving the inverse kinematic problem for robotic manipulators," <u>Prep. of the Sixth CISM-IFToMM Ro.Man.Sy</u>, 82-89.

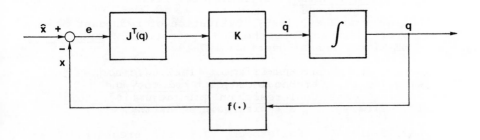

Fig.2.1 The closed-loop scheme of the inverse kinematic solution algorithm.

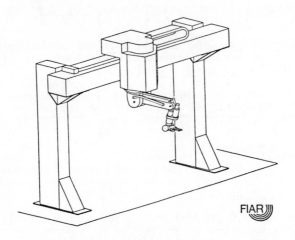

Fig.3.1 The prototype robot DEXTER
(courtesy of FIAR S.p.A.)

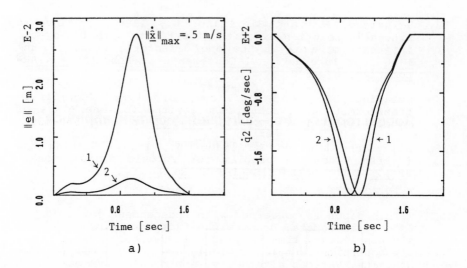

Fig.3.2 a) End-effector tracking error.
        b) Shoulder joint velocity q_2.
           1 - without gain adjusting,
           2 - with gain adjusting.

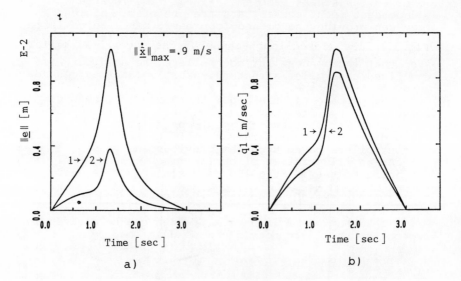

Fig.3.3 a) End-effector tracking error.
        b) Sliding joint velocity q_1.
           1 - without gain adjusting,
           2 - with gain adjusting.

Chapter 17

Robust robot control—a frequency domain approach

S. Engell and A. Kleiner

1.   INTRODUCTION

This paper is concerned with independent joint control
of industrial robots. Despite the fact that a robot is a
highly non-linear, time-varying multivariable system, this
simple control structure is widely and successfully used
in practice. Various alternative approaches have been proposed and investigated, adaptive (self-tuning) control
(e. g. Lelic and Wellstead (1)) and non-linear compensation
(e. g. Freund (2)) being the main directions of research.
As such more sophisticated schemes are of course more demanding in terms of computing power and because of not yet
resolved robustness problems, linear time-invariant (lti)
independent joint controllers are preferable as long as
they can provide the closed-loop performance which is required. The performance of the robot controllers which are
commercially available however is not only limited by the
fact that independent lti controllers are used but also by
the fixed controller structure, usually cascaded controllers of PI-/PD-type.
The purpose of this paper therefore is two-fold:
1) to provide a framework for the analysis of the potential of lti independent joint controllers
2) to discuss control structures of lti independent joint
   controllers based upon the previous analysis.
As an illustration of our approach, the control of
one joint of the IR Kuka 160 is discussed in detail.

2.   INDEPENDENT JOINT CONTROL AS A ROBUST CONTROL PROBLEM

The dynamics of industrial robots can be divided into
two parts: the rigid body dynamics described by

$$\underline{f} = \underline{M}(\underline{q})\underline{\ddot{q}} + \underline{H}\underline{\dot{q}} + \underline{c}(\underline{q},\underline{\dot{q}}) + \underline{g}(\underline{q}) \tag{2.1}$$

where $\underline{f}$ is the (nxl) vector of actuated generalized forces,
$\underline{M}(\underline{q})$ is the position-dependent (nxn) matrix of inertia, $\underline{H}$
is the (nxn) diagonal viscous friction matrix, $\underline{c}(\underline{q}, \underline{\dot{q}})$ is
the vector of Coriolis and centrifugal forces and $\underline{g}(\underline{q})$
describes the influence of gravity and compensation of

gravity. $\underline{q}$ denotes the vector of actual joint positions.

In the case of robots with gears, the actuated forces cannot be commanded directly but only via elastic components. Therefore in the frequency domain, we have ($i=1..n$)

$$f_i(s) = G_{Ei}^r(s) \cdot [q_{Ii}(s) - r_i q_i(s)] = G_{Ei}^r(s) \cdot \Delta q_i(s) \quad (2.2)$$

where q_{Ii} denotes the i-th motor shaft position and r_i the i-th gear ratio. In addition, non-linear effects may be present in the transmission from $\Delta q_i(t)$ to $f_i(t)$. The inner position q_{Ii} in turn is determined by the motor dynamics which can be modelled as

$$\dot{q}_{Ii}(s) = G_M^r(s) [K_C \cdot i_i(s) - \frac{1}{r_i} \cdot f_i(s)], \quad (2.3)$$

where $i_i(s)$ is the i-th actuated motor current.

Local linearization of the nonlinear equation (2.1) around the desired position $\underline{q}_d$ yields

$$\delta f_i(t) = J_i(\underline{q}_d, M_L)\delta q_i(t) + H_{ii}\delta \ddot{q}_i(t) + \underline{G}_i(\underline{q}_d, M_L)\delta \dot{q}_i(t) + d_i(t), \quad (2.4)$$

where M_L represents the payload and $d_i(t)$ comprises the external forces acting on joint i which result from the coupling terms in (2.1) and changes in M_L. So locally, the i-th joint dynamics can be represented in the block diagram shown below.

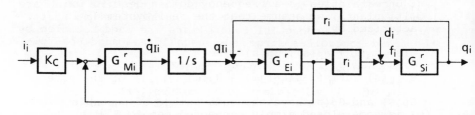

Figure 1: Structure of joint dynamics for linearized model

The transfer function $G_{Si}(s)$ which results from (2.4) varies with $\underline{q}_d$ and M_L, and if the controller design is based on a fixed model of the system dynamics, deviations of the real transfer function $G_{Si}^r(s)$ from the nominal function $G_{Si}(s)$ are to be expected. Also, the systems G_{Mi}^r and G_{Ei}^r are only known up to a certain accuracy. G_{Ei}^r may also vary with the position of the robot and with time (due to mechanical deterioration).

Therefore, any fixed linear time-invariant joint controller must have pronounced robustness properties. We shall discuss in the next section how this robustness requirement limits the attainable closed-loop performance.

It is well known, that satisfactory control for all lti plants G_{Mi}^r, G_{Ei}^r, G_{Si}^r is necessary but not sufficient for robust control of the non-linear system (2.1) together with (2.2) and (2.3). However, as long as the variations of the closed-loop dynamics are not too large, in most cases controllers which yield satisfactory control of a family of lti plant also achieve good control for time-varying or nonlinear plants which are locally in this family.

3. MODELLING OF PLANT UNCERTAINTIES

We use the term plant uncertainty for the characterization of any deviation of the real behaviour of a real dynamical system G_p^r from some nominal description G_p. Such deviations may be <u>structured</u>, e. g. because a lti model is used for a non-linear time-varying plant with well-known dynamic behaviour. This situation is present in the case of eq (2.4) where the parameters J_i and G_i vary with the position and the payload of the robot.

A more general situation is that the real plant dynamics can only be assumed to be "close" to the nominal dynamics in the sense that a compact neighbourhood around the nominal model contains all possible real dynamics. Such unstructured uncertainties are encountered in any real control problem.

The most appropriate characterization of unstructured plant uncertainties is a frequency domain description. For SISO lti plants, we assume that the real system G_p^r is characterized by

$$G_p^r(s) = (1 + \varepsilon_p(s)) \, G_p(s) \qquad (3.1)$$

where $\varepsilon_p(s)$ satisfies $\varepsilon_p(j\omega) \in R_p(\omega)$ $\qquad (3.2)$

and $G_p^r(s)$ and $G_p(s)$ have the same <u>number</u> of unstable poles. $R_p(\omega)$ is some closed simply connected set in $\mathbb{C}$ with the property that any straight line from the origin to the boundary of R_p is completely in R_p

Usually, $R_p(\omega)$ is assumed to be a circle around the origin with radius $\ell_p(\omega)$ (cf. Doyle and Stein (3), Chen and Desoer (4)). $\varepsilon_p(j\omega)$ then satisfies

$$|\varepsilon_p(j\omega)| \leq \ell_p(\omega). \qquad (3.3)$$

The sets (3.3) are very convenient to analyse, but as any phase information is disregarded, they may lead to overly conservative results. If the analysis of plant uncertainties is based on frequency response measurements for various operating points, a natural bound on G_p^r is given by the gain and phase variations (compare Engell (5)).

4.   PERFORMANCE BOUNDS

4.1 Unity feedback

Let

$$S(s) = [1 + G_R(s)\ G_P(s)]^{-1} \tag{4.1}$$

be the nominal disturbance-to-output transfer function of a feedback loop with plant G_P and controller G_R, and let

$$T(s) = 1-S(s) \tag{4.2}$$

be the complementary sensitivity function.

Theorem (5)

Let the real plant be described by (3.1), (3.2) and let $d(\omega)$ denote the (minimal) distance from the point $-T^{-1}(j\omega)$ to the set R_P. If the point is inside $R_P(\omega)$, $d(\omega)=0$. Then the closed-loop system is stable iff the nominal cl-system is stable and

$$\inf_{\omega} d(\omega) > 0. \tag{4.3}$$

If $R_P(\omega)$ is the circle (3.3), (4.3) simplifies to (3, 4)

$$\sup_{\omega} |T(j\omega)\ \ell_p(\omega)| < 1. \tag{4.4}$$

The effect of the restriction (4.4) on the attainable closed-loop performance has been studied extensively in (Engell (6, 7)). From these results, the following conclusion can be drawn:

Performance rule

Assume that (4.4) implies that $|T(j\omega)|$ must be smaller than ε_T for $|\omega|>\omega_z$. Then for not too large maximal values of $|S(j\omega)|$, the bandwidth over which good disturbance rejection can be achieved is approximately given by $\varepsilon_T \cdot \omega_z$ if the plant has no finite poles and zeros with positive real part.

As

$$S^r(j\omega) = S(j\omega)/(1+\varepsilon_p(j\omega)T(j\omega)),$$

robust performance can be achieved in principle as long as $|\varepsilon_P(j\omega)|$ is considerably smaller than 1. In situations where good tracking is desired, as e. g. in robot control, however, it is in addition necessary to bound the peak value of $T^r(j\omega)$ in order to guarantee small overshoot also

for the real control loop. $d(\omega)$ should be larger than
$1\ T_{max}$. This restricts the range where $|T(j\omega)|$ peaks to
those frequencies where $|\varepsilon_P(j\omega)| < .3$, unless R_P has a special
shape such that the phase information about $\varepsilon_P(j\omega)$ can be
used for a less conservative estimate.

4.2 Cascaded control loops

The standard configuration of robot joint control
loop is a cascaded structure where the inner loop controls
motor shaft velocity and the outer loop controls joint
position using the velocity set point as the actuated input. Such a control structure is depicted in fig. 2.

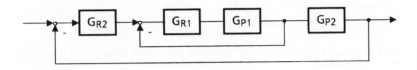

Figure 2: Structure of cascaded control

For this structure,

$$G_P^r(s) = G_{P2}^r(s) \cdot \frac{G_{R1}(s)G_{P1}^r(s)}{1+G_{R1}(s)G_{P1}^r(s)} \quad ,$$

and so approximately

$$\varepsilon_P(s) = \varepsilon_{P2}(s) + \varepsilon_{P1}(s)\,\frac{S_1(s)}{1+\varepsilon_{P1}(s)T_1(s)} \quad .$$

As long as the uncertainty in the inner loop is not
too large, it can be reduced by the use of feedback and
the uncertainty in the outer loop is smaller than it would
be without cascaded control.

Cascaded control therefore is usefull for the reduction of the effect of uncertainties on the closed-loop
performance, iff, at a particular frequency,
(a) the uncertainty of the overall plant $G_{P1}^r G_{P2}^r$ is larger
   than the uncertainty of G_{P2}^r
(b) the relative uncertainty in the inner loop is small
   enough to allow an effective reduction by feedback.

Thus cascaded control is helpfull in the case of a
series of plants with independent moderate uncertainties,
because it can be used to considerably reduce the individual dynamic variations. Conversely, cascaded control offers
no advantage if the individual uncertainties are relatively
large or have such a structure that they cancel each other.

5.   NOMINAL MODEL AND UNCERTAINTIES FOR ROBOT JOINT DYNAMICS

5.1 General approach

In order to simplify the notation, we suppress the index "i" in this section, all variables involved are those associated with one particular joint of the robot. Following the standard approach, (e. g. Kuntze (8)), we chose the following nominal model (compare fig. 1):

$$G_M(s) = \frac{1}{K_D + J_I s} \tag{5.1}$$

$$G_E(s) = C + Ds \tag{5.2}$$

$$G_S(s) = \frac{1}{Js^2} . \tag{5.3}$$

H_{ii} and G_i in (2.4) are neglected because their influence is rather small.

The variation of J_i with the position and the load of the robot causes a structured perturbation of $G_S(s)$. Unstructured uncertainties must be assumed to occur in all three transfer functions, but they can be expected to be most pronounced in $G_E(s)$ because a complex elastic system (a harmonic drive gear and an elastic belt in our case) is modelled by a simple damped spring.

For control purposes, the transfer functions from input current to motor shaft velocity, $\tilde{G}_M(s)$ and from motor shaft position to outer position, $G_Q(s)$ are important. They result as

$$\tilde{G}_M^r(s) = K_C \cdot \frac{s(1+r^2 G_E^r(s) G_S^r(s)) G_M^r(s)}{s(1+r^2 G_E^r(s) G_S^r(s)) + G_M^r(s) G_E^r(s)} \tag{5.4}$$

$$G_Q^r(s) = \frac{r G_E^r(s) G_S^r(s)}{1+r^2 G_E^r(s) G_S^r(s)} . \tag{5.5}$$

The overall transfer function from input current to outer position, $G_P(s)$, is given by

$$G_P^r(s) = \frac{1}{s} \cdot G_Q^r(s) \tilde{G}_M^r(s) . \tag{5.6}$$

The nominal transfer functions $\tilde{G}_M(s)$, $G_Q(s)$, $G_P(s)$ result from the insertion of (5.1)-(5.3) into (5.4)-(5.6). One realizes, that for $\tilde{G}_M$ and G_Q, and hence for G_P, both structured and unstructured uncertainties are present, and that the (uncertain) denominator of $G_Q(s)$ cancels in the calculation of $G_P^r(s)$.

5.2 Analysis of a particular robot

In our analysis of the control of joint 2 of the IR Kuka 160 (see fig. 3), we proceeded as follows:

1. Theoretical analysis of the dependency of J_i on the posi-tion and the load of the robot. The result was (Kleiner (9))

$$J(q_3, M_L) = 348,2 - 25,3\cos q_3 + (2,40+1,91\cos q_3) \cdot M_L, (5.7)$$

   where M_L is the payload (in kg) which varies in the range of 0-45 kg. (5.7) shows that the effect of the position alone is rather small (less than 8 %), but the combined effect of payload and position leads to a 40 % increase of J compared to the minimal value.
2. Calculation of the frequency responses $\tilde{G}_M(j\omega)$ and $\tilde{G}_Q(j\omega)$ from experimental data for a nominal position ($q_2 = 90°$, $q_3 = 0°$) and no payload. The nominal transfer functions $G_M(s)$ and $G_E(s)$ were then obtained by least-squares approximation of these frequency responses. Measured frequency responses and approximations are shown in fig. 4a, b.
3. Calculation of frequency responses $\tilde{G}_M(j\omega)$ and $\tilde{G}_Q(j\omega)$ from experimental data for various positions q_2, q_3 and no load. Because of the result of step 1, the observed changes of the dynamic behaviour can be mainly attrib-uted to the uncertainty in G_E and be modelled as un-structured uncertainties. The main results of this step were:
   - the uncertainty for frequencies up to 20 rad/s is below 10 %, i. e. the model provides a good basis for controller design
   - large uncertainties were observed in the frequency range of 20-30 rad/s, the moduli of $\tilde{\epsilon}_M(s)$, $\epsilon_Q(s)$ and $\epsilon_P(s)$ all exceed 1 and the phase decreases up to 120° compared to the nominal model
   - around the resonance frequency of $G_Q(s)$ (40-50 rad/s), large perturbations are observed in G_M and G_Q but much smaller perturbations result for $G_P(s)$ (compare fig. 5).
4. Analysis of the effect of the payload based on the nomi-nal model and eq (5.7) with the constant term adjusted according to the result of step 2. The result of this analysis was that the resonance frequency and damping of $G_Q(s)$ vary about 20 %. However, the variation of $G_P(s)$ is much less pronounced.

6.   CONCLUSIONS ON CONTROL STRUCTURE AND PERFORMANCE

From the performance rule in section 4 and the results of the identification, we can conclude that the attainable bandwidth for the position control loop is limited to 10-15 rad/s. This is due to the identified unstructured per-

turbations around ω = 25 which were not predicted by the
theoretical model. In contrast, the structured uncertainties caused by the variation of the inertia are much less
pronounced. If only these perturbations were present, a
bandwidth of at least 20 rad/s could be achieved.

As far as the reduction of uncertainties is concerned,
direct control of the outer position should be preferred.
The reason for this is that the uncertainties in the inner
and the outer loop are more difficult to control than the
uncertainty of the overall loop. If for reasons other than
reduction of uncertainty a cascaded structure is preferred,
the bandwidth of the inner loop should not exceed 10-12
rad/s.

In figure 6, the closed-loop frequency response for
the nominal case (q_2=90°, q_3=0°, M_L=45 kg) and for a situation where large deviations are present (q_2=60°, q_3=-
130°, M_L=0) are shown. The controller used is of order
three and consists of a lead compensator and a term with
conjugate poles and zeros. Gain crossover is at 10 rad/s
and phase margin $\approx$ 70°.

7.   SUMMARY

We presented an approach to the analysis of independent joint control using frequency response methods. It was
shown how performance bounds for the individual loops can
be obtained. For the example which we studied, the unstructured uncertainties identified are more important than the
structured ones (variation of the inertia). This points to
the fact that the frequency domain identification of the
dynamics including plant uncertainties is a valuable tool
for the choice and design of a control structure.

The robustness analysis presented here is of course
not a comprehensive answer to the question whether the
performance of the real system will be satisfactory. This
also depends on the strength of the disturbances which
enter in (2.4) and on the non-linear effects as friction
and backlash.

8.   ACKNOWLEDGEMENTS

The authors are gratefull for stimulating discussions
with H.-B. Kuntze, A. Jacubasch and W. Schill, all at IITB
Karlsruhe.

9.   REFERENCES

(1) Lelic, A.M., and Wellstead, P.E., *Int. J. Control*, 46,
    569-601.
(2) Freund, E., Proc. IFAC Symposium on Multivariable
    Technological Systems, 1977, Pergamon Press.
(3) Doyle, J.C., and Stein, G., *IEEE Tr. on AC*, 26, 4-16.
(4) Chen, M.J., and Desoer, C.A., *Int. J. Control*, 35,
    255-267.

(5) Engell, S., and Kleiner, A., Workshop Kinematic and
    Dynamic Issues in Sensor Based Control, Castelvecchio,
    1987, Proc. to be published by Springer Verlag.
(6) Engell, S., Grenzen der erreichbaren Regelgüte, Habi-
    litationsschrift University of Duisburg, 1987, to be
    published by Springer Verlag.
(7) Engell, S., Proc. 10th IFAC World Congress, Munich,
    1987.
(8) Kuntze, H.-B., IFAC Symposium SYROCO'85, Barcelona,
    1985.
(9) Kleiner, A., Diplom-Ingenieur thesis, University of
    Karlsruhe, 1987.

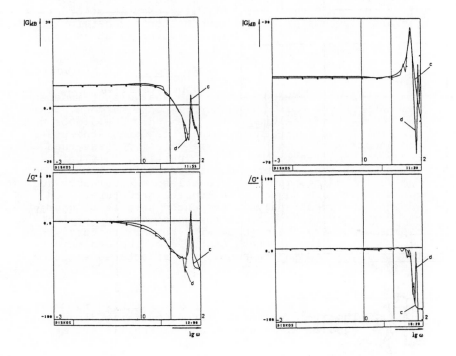

Figure 4: Measured (d) and approximated frequency responses
          Left: $\tilde{G}_M$ Right: G_Q

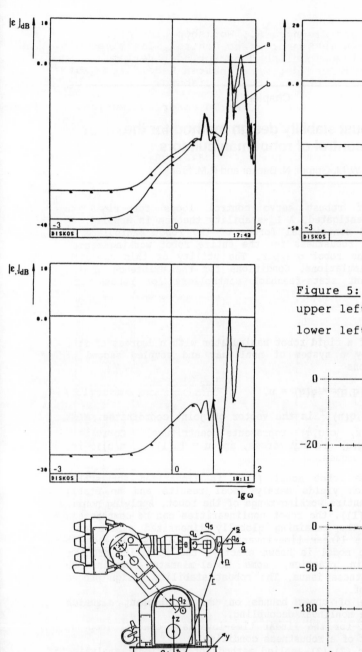

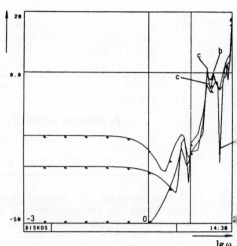

Figure 5: $|\varepsilon|$ for various case
upper left: $\varepsilon\tilde{M}$, upper right: ε
lower left: ε_p

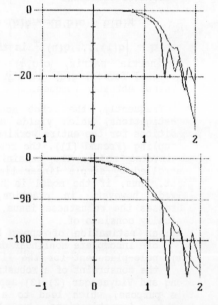

Figure 3: The IR KUKA 160/45

Figure 6: Closed-loop
frequency responses,
designed and real

Chapter 18

A robust stability design method for the control of robot manipulators

W. M. Grimm, N. Becker and P. M. Frank

The design of robust servo control loops for rigid robot manipulators is investigated. A L_∞-stability theorem is applied, which takes Coriolis- and centripetal forces of the robot manipulator into account, and yields robustness for the entire robot working range and varying loads of the robot gripper. The utility of this approach is demonstrated via simulations. Conditions for the existence of robust linear time-invariant state-feedback controllers for robots can be obtained.

1. INTRODUCTION

The dynamics of a rigid robot manipulator with n degrees of freedom can be described by a system of nonlinear and coupled second order differential equations

$$M(q)\ddot{q} + c(q,\dot{q}) + g(q) = u, \tag{1.1}$$

where $q = \left[q(1),\ldots,q(n)\right]^T$ is the vector of joint coordinates. $M(q)$ is the inertia matrix, $c(q,\dot{q})$ represents centrifugal, Coriolis and frictional forces, $g(q)$ gravity forces, and $u = [u(1),\ldots,u(n)]^T$ is the vector of actuator torques.

Frequently, the robot model (1.1) is linearized for stability investigations, which yields merely local results and no stability conditions for the entire working range of the robot. Applying nonlinear decoupling (Freund (1)), the robot nonlinearities can be compensated by an inverse model. The remaining globally linearized system can be controlled by simple linear time-invariant controllers for each robot joint. Thus, if the model is known exactly, the stability problem is solved; however, in practice, some model-mismatch remains which addresses the robustness issue. The robust stability design procedure basically consists of

(i) an estimation of upper bounds on the model-plant mismatch of of incomplete nonlinear decoupling,

(ii) pole-placement for the linear time-invariant control loop under the constraint of a robustness condition.

Spong and Vidyasagar (2)-(3) applied methods of functional analysis for this purpose, which lead to a clearly laid out stability analysis. Further investigations by Becker and Grimm (4) showed that modifications of this method are *necessary* and extensions are possible. The resulting sufficient L_2- and L_∞-stability conditions in (4) can be used for the design of non-adaptive robust control loops. This short paper focuses on the applicability of the L_∞-stability approach for the practically important linear time-invariant controllers, which may be viewed as a worst-case situation for the robustness analysis.

In Section 2, a L_∞-stability condition for a class of nonlinear time-variant control systems is stated, and is subsequently applied to robot manipulators. Section 3 summarizes some techniques for estimating bounds on robot nonlinearities. In Section 4, a robust linear time-invariant state-feedback controller for the robot VW-G60 is studied as an example, and simulations are carried out which demonstrate the performance of the proposed controller. Concluding remarks are given in Section 5.

2. ROBUST ROBOT MANIPULATOR CONTROL LOOPS

2.1 The Stability Theorem

A class of nonlinear time-variant control loops is considered which can be represented by Fig. 2.1, where $\eta(e_1,e_2,v,f_1(t),\ldots,f_\mu(t))$ is introduced which consists of all static nonlinearities of plant and controller and acts as a perturbation on the linear closed-loop feedback

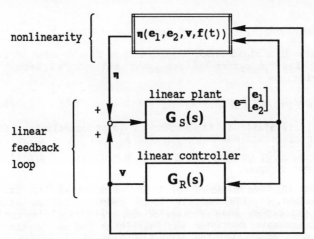

Fig. 2.1 Representation of the nonlinear time-variant
control loop for stability investigations.

system. $f_i(t)$ $(i = 0,1,\ldots,\mu)$ are known time functions. By partitioning e, i.e. considering more measurable system states for the stability analysis, the dependence of the nonlinearity η on these states can be taken into account in more detail, which results in less conservative nonlinearity bounds. Let e_j denote the output of the transfer matrix $G_{Sj}(s)$. The input-output relations of the closed loop in Fig. 2.1 can be described by

$$e_j(t) = G_j(t) * \eta(t) + i_{ej}(t), \quad (j=1,2), \qquad (2.1)$$

$$v(t) = G_3(t) * \eta(t) + i_v(t), \qquad (2.2)$$

where $G_j(s) = G_{Sj}(s)\Phi(s)$, $G_3(s) = \Phi(s)G_R(s)G_S(s)$,

and $\Phi(s) = (I-G_R(s)G_S(s))^{-1}$. $i_{ej}(t)$ and $i_v(t)$ denote initial condition responses, and the asterisk in (2.1)-(2.2) the convolution operator. In principle, further partitioning can be considerd for e as well as for v. $G_R(s)$ is chosen such that $G_i(s)$ (i=1,...,3) in (2.1)-(2.2) are stable transfer matrices. Then, the induced L_∞-operator norms of $G_i(s)$ exist which are denoted by β_i in the sequel. Further, let the nonlinearity **η** in Fig. 2.1 estimated by the sum of Euclidean norms

$$\|\mathbf{η}\| \leqslant k_0\|\mathbf{e}_2\|^2 + k_1\|\mathbf{e}_1\| + k_2\|\mathbf{e}_2\| + k_3\|\mathbf{v}\| + \sum_{\ell=1}^{\mu} k_{f\ell}\|f_\ell\|^{m_\ell} + k_4, \qquad (2.3)$$

where k_i (i=1,...,4) and $k_{f\ell}$ (ℓ=1,...,μ) are non-negative finite constants. Further, consider the input functions $f_\ell(t)$ (ℓ=0,1,...,μ) to be bounded. The form of the estimate (2.3) is one of several possibilities. This particular form is motivated by an application to robot manipulator control.

Theorem 2.1: If $\|\mathbf{e}_2(t=0)\|$ is sufficiently small then the closed-loop nonlinear time-varying feedback system in Fig. 2.1 is
(i) <u>weak</u> L_∞-stable if

$$1 - \beta_1 k_1 - \beta_2\left[k_2+2(k_0 d)^{1/2}\right] - \beta_3 k_3 > 0 , \qquad (2.4)$$

where

$$d = \sum_{\ell=1}^{\mu} k_{f\ell}\|f_\ell\|_{max}^{m_\ell} + k_1\|i_{e1}\|_{max} +$$

$$+ (1-\beta_1 k_1 - \beta_3 k_3)/\beta_2\|i_{e2}\|_{max} + k_3\|i_v\|_{max} + k_4 \qquad (2.5)$$

(ii) L_∞-stable if (2.4) holds and $k_4 = 0$ in (2.3) and (2.5).
<u>Proof:</u> Analogous to Becker and Grimm (5) for $k_2 \neq 0$.

2.2 Application to Robot Manipulator Control Loops

As outlined in the introduction, the robustness analysis relies upon an estimation of the model-plant mismatch of a compensating controller. In principle, Theorem 2.1 can be applied to robot control loops with for example complete or incomplete nonlinear decoupling (Freund, (1)), gain scheduling and static decoupling (Williams (6)), or pure linear control when the nonlinearity **η** describes plant nonlinearities on their own. The theoretically best controller which is nonlinear decoupling with an exact model would yield **η** = **0**; however, it would be highly nonlinear and complex and, therefore, not of practical value for fast real time implementations.

It is desired that the robot manipulator follows a prescribed reference trajectory $q_d(t)=[q_d(1),...,q_d(n)]^T$ in the joint coordinate system, where $\partial q_d/\partial t(t=0)=0$, and $q_d(t\to\infty):=q_{d\infty}$ is assumed to be constant. The purpose of control is, firstly, to keep the tracking errors

$$e_1 := q - q_d , \; e_2 := \dot{q} - \dot{q}_d \qquad (2.6)$$

dynamically small, and secondly, to achieve no offset at steady state. The first objective requires sufficiently smooth transition curves of the desired reference trajectory $q_d(i)$, $(i=1,\ldots,n)$. The second requires integral action of the employed controller. Let the actuator torques u be evaluated from the control law

$$u = \hat{M}[\ddot{q}_d + v] + \hat{c} + \hat{g} \qquad (2.7)$$

where ^ denotes controller matrices and vectors which can represent linear or nonlinear, time-variant or time-invariant model quantities corresponding to (1.1). The vector v is the output of an integral linear time-invariant feedback controller which has to be specified by pole-placement. The feedforward action in (2.7), i.e. the second derivative of $q_d(t)$, may be dropped; however, it is important for a good servo response.

In the sequel let us focus on the applicability of the L_∞-approach to the special case of a linear controller ($\hat{M}$ = const., $\hat{c} = \hat{g} = 0$). Then, the nonlinearity becomes (5)

$$\eta = [M^{-1}(\bar{q})\hat{M}-I] \; [v-\hat{M}^{-1}g(q_{dr}) + \ddot{q}_d] +$$
$$+ M^{-1}(\bar{q}) \, [g(q_{dr}) - g(\bar{q}) - c(\bar{q},\dot{q})], \qquad (2.8)$$

where the lower index r denotes a constant reference value, and $\bar{q} = q - q_{dr}$ is introduced due to a *necessary* reformulation of the control loops (4),(5). A comparison of the estimate of η from (2.8) and (2.3) leads to the "k-values" which are needed for Theorem 2.1. The control loop nonlinearities can be estimated in the following manner:

$$\|M^{-1}(\bar{q})\| \leq a_1 \qquad (2.9)$$

$$\|M^{-1}(\bar{q})\hat{M}-I\| \leq a_2 \qquad (2.10)$$

$$\|g(\bar{q}) - g(q_{dr})\| \leq b_1\|\bar{q}\| \qquad (2.11)$$

$$\|c(\bar{q},\dot{q})\| \leq b_0^2 \; \|\dot{q}\|^2 + b_2\|\dot{q}\| + b_3 . \qquad (2.12)$$

The estimate (2.11) is due to gravitational forces. In (2.12) Coriolis and centripetal forces are estimatd by b_0, viscous friction by b_2, and, finally, Coulomb friction by b_3.

From (2.3) and (2.9)-(2.12) it follows that

$$k_0 = 2a_1b_0, \; k_1 = a_1b_1, \; k_2 = a_1b_2, \; k_3 = a_2, \; k_4 = a_1b_3,$$

$$\sum_{\ell=1}^{\mu} k_{f_\ell}\|f_\ell\|^{m_\ell} = 2a_1b_0\|\dot{q}_d\|^2 + a_1b_2\|\dot{q}_d\| + a_1b_1\|\bar{q}_d\| + a_2\|\ddot{q}_d\| \qquad (2.13)$$

and Theorem 2.1 can be applied for the robot manipulator case. In order to simplify (2.5), $q_{dr} = q_d(t=0)$ is chosen which leads to $i_{e1}(t) = i_{e2}(t) = i_v(t) = 0$. Thus, (2.5) and (2.13) yields

$$d = k_0\|\dot{q}_d\|_{max}^2 + k_1\|q_d-q_d(0)\|_{max} + k_2\|\dot{q}_d\|_{max} + k_3\|\ddot{q}_d\|_{max} + k_4$$

$$(2.14)$$

While the "β-values" in (2.4) depend entirely on pole-placement of the linear control loop ($\eta=0$), the "k-values" in (2.4) and (2.14) can be viewed as properties of a specific robot manipulator. Let us briefly sketch the evaluation of the latter values in the next paragraph.

3. BOUNDS ON ROBOT MANIPULATOR NONLINEARITIES

The practically important case of linear time-invariant controllers is discussed. This case can be viewed to give the worst k-values for the robustness analysis since a nonlinear controller should always be designed such that it compensates for plant nonlinearities in some optimal manner.

The constant a_1 in (2.9) can be evaluated easily with a search program over all robot configurations and loads. The constant a_2 depends on the choice of $\hat{M}$. According to (2.7) this choice determines already one part of the controller. For the ideal case of a complete nonlinear decoupling, $a_2=0$ would be obtained; however, for $\hat{M}$ = constant, a_2 will be less than unity for a meaningful choice. It is evident that the smallest value of a_2 is desirable. This can be achieved by computing

$$a_2 = \min_{\hat{M}} \max_{q,m_L} \|\hat{M}^{-1}(q,m_L)\hat{M} - I\|, \qquad (3.1)$$

where $m_L \in [0, m_{Lmax}]$ is the load mass. Since β_3 is always greater than one $a_2 < 1$ is *necessary* for the applicability of Theorem 2.1. $\beta_3 > 1$ is due to the integral character of the plant which results in $G_3(j\omega=0) = -I$. From experience, $a_2 < 1$ can be easily achieved for robots with gears since motor inertia multiplied by the square of the inverse gear ratio dominates robot inertia. More effort may be necessary for determining a small value a_2 when direct drive robots are controlled. $a_2 < 1$ is *necessary* and *sufficient* for the existence of a linear time-invariant controller which fulfils the L_∞-stability condition.

The evaluation of b_0 and b_1 is demonstrated for the first three degrees of freedom in (5) using the robot VW-G60 as an example. Its complete model can be gathered from Klein (7). Finally, the constants b_2 and b_3 are the friction coefficients.

The proposed procedure for estimating robot nonlinearities in (5) can be carried over to many types of robot manipulators. The computation of the bounds seems to be expensive; however, it has to be done just once for each robot manipulator.

4. A L$_\infty$-ROBUST LINEAR STATE FEEDBACK CONTROLLER FOR THE VW-G60 ROBOT

The design objective for a linear time-invariant controller is to meet the L_∞-robustness constraint (2.4). Therefore, small β-values are desired. Since the L_∞-approach yields explicit bounds on the tracking error as a function of model uncertainty, small β-values are desired for good performance as well. While $\beta_3 > 1$ for any linear controller, β_1 and β_2 can be made arbitrarily small by shifting at least one closed-loop pole towards $-\infty$. However, controller gains become very large under these circumstances. Thus, there exists a trade-off between robustness and sensitivity of the closed loop with respect to measurement noise.

The measurement of joint angle position and velocity is typically accomplished with a resolution of 16 bit for 360 degrees. Then, it can be shown that using a state feedback controller with Davison compensator, a closed-loop single pole at $\lambda > -27.7$ and a double pole at

$\lambda/4$ yields actuator torques which do not exceed 30% of the momentum to overcome static friction for the VW-G60 robot.

The design procedure is as follows:

(a) Determine $\hat{M}$ and a_2 from (3.1). Note that $a_2 < 1$ has to be achieved for the existence of a robust controller.

(b) Obtain the upper bounds on the nonlinearities in (2.9)-(2.12), and evaluate k_i (i=0,...,4) from (2.13) which are properties of the robot manipulator mechanics. Further, determine d from (2.14) which depends on the robot working range, speed and acceleration as well.

(c) Design $G_R(s)$ such that the L_∞-stability condition (2.4) is fulfilled. For state-feedback controllers this can be done conveniently by using design graphs for the β-values which are available in (5).

(d) The former step gives a constraint on pole-placement for robustness. Now check whether the desired measurement noise limitations are violated.

As an example, the path contour shown in Fig. 4.1(a) and 4.2(a) is considered which is known to be used in industry for checking position accuracy. It consists basically of two circles with radii of 15 mm and 40 mm which are to be tracked by the robot manipulator in the horizontal X-Y-plane. Arrows in Fig. 4.1(a) and 4.2(a) indicate the direction of the movement. The robot starts moving with zero velocity at point A, which corresponds to $q(1) = 0°$, $q(2) = 30°$, $q(3) = 150°$ in the joint coordinate system, and ends in point A with zero velocity.

Linear SISO controllers for each robot joint are obtained if a diagonal matrix is chosen for the constant matrix $\hat{M}$. For the VW-G60 robot step (a) yields $\hat{M}(1,1) = 1325$, $\hat{M}(2,2) = 1206$, $\hat{M}(3,3) = 568$ in kgm^2 and $a_2 = 0.27 < 1$. Step (b) gives $k_0 = 3.24/rad$, $k_1 = 14.9/s^2$, $k_2 = 1.76/s$, $k_3 = 0.27$ and $k_4 = 0.88rad/s^2$. A triangular acceleration profile is chosen which is characterized by

$$\|\dot{\ddot{q}}_d\|_{max} = 3.1 \text{ rad/s}^2, \quad \|\ddot{q}_d\|_{max} = 0.35 \text{ rad/s}, \quad \|\bar{q}_d\|_{max} = 0.11 \text{ rad} \qquad (5.1)$$

Then, d = 3.0 rad/s^2 is obtained from (2.14). If step (c) is carried out with a double pole at $\lambda/4$ the design graphs in (5) yield $\lambda > 13.6$ which is below the desired measurement noise limitation. Thus, robust L_∞-stability can be achieved. It should be noted that robust L_2-stability can only be shown when Coriolis, centripetal and frictional forces are neglected (5).

The performance is demonstrated via simulations, where measurement noise and friction are taken into account. Fig. 4.1(b) shows the positon accuracy for an triangular acceleration profile which fulfils (5.1) and causes an average speed of 0.23 m/s. According to Fig. 4.1(c) the actuator torque constraint of 4875 Nm (7) is violated. Thus, the velocity had to be reduced for the small circle which is shown in Fig. 4.2. At the cost of a slightly slower motion of 0.2 m/s in average, the performance could be improved significantly. The maximum tracking error does not exceed 1 mm as shown in Fig. 4.2(b).

5. CONCLUSION

The current industrial philosophy of robot arm control system design is to treat each joint of the robot arm individually as a simple servomechanism with for example cascade controllers. This approach is inadequate for a big class of robot manipulators because the motion

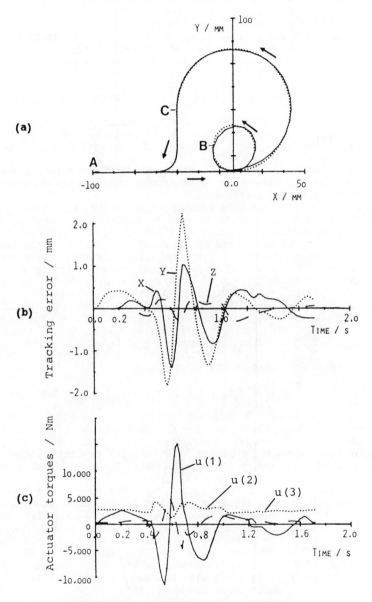

Fig. 4.1 Performance for an triangular acceleration profile,
where A(t=0.0s), B(t=0.67s), C(t=1.28s).
(a) Position trajectory in the X-Y-plane
    (..... desired path,————— robot VW-G60),
(b) Corresponding tracking error,
(c) Corresponding actuator torques.

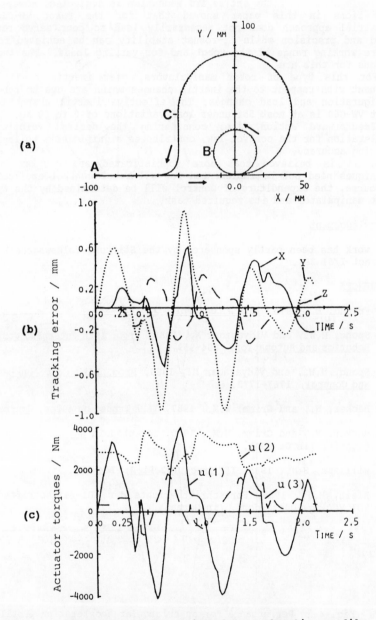

Fig. 4.2 Performance for an triangular acceleration profile,
where A(t=0.0s), B(t=0.79s), C(t=1.61s).
(a) Position trajectory in the X-Y-plane
    (..... desired path,————— robot VW-G60),
(b) Corresponding tracking error,
(c) Corresponding actuator torques.

and configuration of the entire arm mechanism is neglected. However, the simulations in this paper showed that for the robot VW-G60 the industrial approach does not necessarily lead to poor servo response speed and precision while L_∞-robust stability can be achieved for the entire working range of the robot and for varying loads. The two main reasons for this are

1) For this type of robot manipulators, gear inertia tends to be dominant with respect to the inertia changes which are due to robot arm configuration and load changes. The effective inertia change of the robot VW-G60 is at most 55% under load variations of 0 to 60 kg.

2) Feedforward action, i.e. considering the desired velocity and acceleration for the control law, contributes significantly to improving tracking accuracy.

Thus, it is believed that more sophisticated and complex control techniques need not be implemented for all robot control applications. Of course, the expenditure of control will be determined by the type of robot manipulator and its required task.

Acknowledgement

This work has been partly sponsered by the Stiftung Volkswagenerk under Contract I/61 394.

REFERENCES

1. Freund, E., 1975, Int. J. Control, 21, 443-450.

2. Spong, M.W., and Vidyasagar M., 1985, Proc. IEEE Internat. Conf. on Robotics and Automation, 954-959.

3. Spong, M.W., and Vidyasagar M., 1985, Proc. 24th Conf. on Decision and Control, 1767-1772.

4. Becker, N., and Grimm, W.M., 1987, IEEE Trans. on Autom. Contr.,32.

5. Becker, N., and Grimm, W.M., 1988, accepted in Automatisierungs-technik, (in German).

6. Williams, S.J., 1985, IEE Proc.,132,Pt.D, 144-150.

7. Klein, H.J., 1984, Fortschrittberichte der VDI-Zeitschriften, 81, VDI Verlag, Düsseldorf, (in German).

Chapter 19

Robust decentralised position control of industrial robots with elasticities and coulomb friction

P. C. Müller and J. Ackermann

1. INTRODUCTION

The application of industrial robots to advanced manufacturing tasks requires highly accurate position and/or
force control. Actual limitations to these requirements are
mainly caused by elasticity, Coulomb friction and backlash
in the system. Conventional control algorithms do not consider these effects and can hardly provide sufficient control
accuracy or yield limit cycles. In this paper the common mul-
tibody approach of models of industrial robots is extended
to derive more realistic models including elasticity of both
joints and links, Coulomb friction and backlash within the
electric drive. Then in several steps an algorithm for a
highly accurate position control is developed based on (i)
multi-layer control of each robot joint to damp quickly the
elastic motions "separately" from the rigid-body motion of
the robot, (ii) compensation of Coulomb friction and backlash by the method of disturbance rejection control, and (iii)
hierarchical decentralized control of the multi-axes robot
including robust control to coordinate the robot arm motions.

2. SINGLE AXIS CONTROL

Subject of this chapter is the decentralized control of
each axis of a Puma-type robot with three rotational degrees
of freedom for the main motions. In the contrary to the common rigid body approach, here the torsional behaviour of the
flexible drive, the Coulomb friction of the d.c.-motor and
of the harmonic drive gear in each joint as well as the elasticity of each robot arm are taken into account.

2.1 Modeling

According to Ackermann and Müller (1) a state space model of the dynamics of a single robot axis can be represented by a state vector

$$x^T = [\varphi_2 \ \Delta\varphi_3 \ \Delta\varphi_5 \ v_1 \ \dot{\varphi}_2 \ \Delta\dot{\varphi}_3 \ \Delta\dot{\varphi}_5 \ \dot{v}_1 \ i_M] . \qquad (2.1)$$

The state variables are the motor angle φ_2, the relative gear
angle $\Delta\varphi_3$, the relative shoulder angle $\Delta\varphi_5$, the amplitude
function v_1 of the first elastic (bending) mode of the arm,
and additionally their derivatives, respectively, and lastly

the motor current i_M. If necessary, an extended state vector may also include variables $\Delta\varphi_1$ of the relative tachometer angle and/or the amplitude functions v_i, $i=2,3,\ldots$, of higher elastic (bending and torsional) modes of the arm, but usually these variables can be neglected because they are related to effects of very high frequencies which are usually not of interest.

The state space equation with respect to the state vector (2.1) is given by

$$\dot{x} = A x + b u + G r \qquad (2.2)$$

where the control input u is the voltage of the electric motor and the vector r represents the nonlinearisties (Coulomb friction, backlash) effecting the robot dynamics. The system matrix A and the input matrices b and G are given as follows:

$$
A = \left[
\begin{array}{cccc|cccc|c}
0_{4\times4} & & & & E_{4\times4} & & & & 0_{4\times1} \\
\hline
0 & -\dfrac{c_{23}}{J_2} & 0 & 0 & -\dfrac{b_2}{J_2} & -\dfrac{b_{23}}{J_2} & 0 & 0 & \dfrac{K_M}{J_2} \\
0 & -\dfrac{c_{23}}{J_2}-\dfrac{c_{23}}{J_3} & -\dfrac{c_{45}}{i^2 J_3} & \dfrac{c_{45}}{iJ_3} & 0 & -\dfrac{b_2}{J_2}-\dfrac{b_3}{J_3} & -\dfrac{b_{23}}{J_2}-\dfrac{b_{23}}{J_3} & -\dfrac{b_3}{J_3} & 0 & \dfrac{K_M}{J_2} \\
0 & -\dfrac{\tilde{c}_{31}}{i}-\dfrac{c_{23}}{iJ_2} & \tilde{c}_{31} & -\tilde{c}_{32} & -\dfrac{b_2}{iJ_2} & -\dfrac{b_{23}}{iJ_2} & 0 & -\tilde{b}_{34} & \dfrac{K_M}{iJ_2} \\
0 & \dfrac{\tilde{c}_{41}}{i} & -\tilde{c}_{41} & -\tilde{c}_{42} & 0 & 0 & 0 & \tilde{b}_{44} & 0 \\
\hline
0 & 0 & 0 & 0 & 0 & 0 & 0 & 0 & -\dfrac{1}{T_I}
\end{array}
\right]
$$
,
$$ \qquad (2.3) $$

$$
b^T = \left[
\begin{array}{cccc|cccc}
0 & 0 & 0 & 0 & 0 & 0 & 0 & 0 & \dfrac{K_I}{T_I}
\end{array}
\right],
$$
$$ \qquad (2.4) $$

$$
G^T = \left[
\begin{array}{cccc|cccc}
0 & 0 & 0 & 0 & -\dfrac{1}{J_2} & -\dfrac{1}{J_2} & -\dfrac{1}{iJ_2} & 0 & 0 \\
0 & 0 & 0 & 0 & 0 & \dfrac{1}{J_3} & 0 & 0 & 0
\end{array}
\right].
$$
$$ \qquad (2.5) $$

Here, the J's represent moments of inertia, the c's and b's are stiffness and damping coefficients, $\tilde{c}$'s are normalized stiffnesses, and i is the gear ratio; K_M, K_I are gains and T_I is the time constant of the electric motor.

The measurement vector

$$ y = C x \qquad (2.6) $$

contains measurements of φ_2, $\dot{\varphi}_2$, v, $\dot{v}$ and additionally of the torque y_5 at the output of the gear to decouple the joint from the connecting link:

$$y_5 = \frac{c_{45}}{i^2} (x_2 - i\, x_3).\tag{2.7}$$

The interesting variable z to be regulated describes the position of the end of the robot arm:

$$z = f^T x = \frac{x_1}{i} - x_3 + \frac{x_5}{l}\tag{2.8}$$

where l is the length of the robot arm.

Due to the Coulomb friction in the motor and in the gear stops may appear in the region of the end position of the robot ($x = 0$). Dependent on the control design the stick friction leads to steady-state inaccuracy (PD-control) or to limit cycles about the equilibrium position (PI-control), cf. Müller and Ackermann (2). Additionally, the controllability of the arm vibrations can be lost, because during the stops the friction torque is reacting the motor torque yielding a decoupling of the control loop.

The Coulomb friction depends on velocity, actuating force and normal pressure. Its characteristic may be described by

$$M_F(\dot{\varphi}) = \begin{cases} M_o & \operatorname{sgn}(\dot{\varphi}) , & \dot{\varphi} \neq 0 \\ M_o & \operatorname{sgn}(M_a), & \dot{\varphi} = 0, \ |M_a| > M_o, \\ M_a & , & \dot{\varphi} = 0, \ |M_a| < M_o \end{cases}\tag{2.9}$$

where M_o depends on normal pressure, and M_a is the actuating force. A more detailed model of Coulomb friction was used in (1,2) including a transition between maximal sticking and gliding. However, the design of a compensation control of the Coulomb friction can be also demonstrated by the model (2.9). Neglecting backlash, the two components of the vector r of the nonlinearities of (2.2) are given by

$$r_1 = M_F(\dot{\varphi}_2),\tag{2.10}$$
$$r_2 = M_F(\dot{\varphi}_2 + \Delta\dot{\varphi}_3).\tag{2.11}$$

The modeling of backlash leads to additional terms in (2.10) and (2.11). Because the method of section 2.3 to compensate the friction is applicable in the same manner to compensate the effect of backlash, in the following we confine ourselves to the problem of friction.

2.2   Linear Control

Neglecting the effects of nonlinearities in equation (2.2), i.e. assuming $r = 0$, the design of a linear feedback control can be evaluated by frequency-response methods or by state space methods such as pole placement or linear optimal control with a quadratic performance index.

Dealing with the design of a linear control it can be shown (1,2) that a static output feedback

$$k_x^T x = k_{x1} y_1 + k_{x4} y_2 + k_{x5} y_3 + k_{x8} y_4 \qquad (2.12)$$

with the state gain vector

$$k_x^T = [k_{x1} \ 0 \ 0 \ k_{x4} \ k_{x5} \ 0 \ 0 \ k_{x8} \ 0] \qquad (2.13)$$

of a linear state feedback control yields very good results. The usual PD-control by feeding back φ_2 and $\dot\varphi_2$ is supplemented by a feedback of v_1 and $\dot v_1$ to damp the elastic vibrations of the arm. Because of the different time-horizons of the signals of the motor angle φ_2 and the bending v_1 the feedback (2.12) can be interpreted as a multi-layer control of the single-axis robot to damp quickly the elastic motions (v_1) separately from the rigid-body motion (φ_2) of the robot. It has to be noted that the measurement (2.7) is not used for the linear control (2.12). The determination of the gains k_{x1}, k_{x4}, k_{x5}, k_{x8} has to be arranged with respect of the change of the effective moments of inertia due to the various positions of the robot arms. It is considered in more detail in section 3 on multi-axes control.

2.3   Compensation of Friction

Vibration systems containing Coulomb friction such as the system (2.2) with r ≠ 0 are nonlinear, and therefore the analysis of dynamical behaviour and stability cannot be carried out by means of linear theory. But in early years it was pointed out by Den Hartog (3) that a linearization of the friction charachteristic by the describing function method may be done, if the ratio of friction to actuating force is small, $M_o/M_a \ll 1$. Then the dependence on sticking behaviour may be neglected and Coulomb friction may be considered as an ideal relay. However, the stability of the resting position ($\varphi = 0$) cannot be analyzed using this harmonic linearization method. Near the desired position the actuating force will decrease such that M_o/M_a increases and stops of motion occur whenever $M_a < M_o$. In the case of a single degree-of-freedom vibration system this behaviour was illustrated by Den Hartog (3) and some more detailed calculations were recently published by Marui and Kato (4). But in the case of a more complicated position-controlled mechanical system (2.2 to 2.11) a satisfactory theoretical analysis does not exist. Therefore, simulation studies were performed which are reported in (1). There it is shown that usual control algorithms and also the state output feedback (2.12) lead to steady-state inaccuracy or to a limit cycle behaviour due to the sticking. To avoid these shortcomings a new feedback control is designed to compensate the Coulomb friction torques.

If in the dynamical system (2.2) the vector r is interpreted as vector of external disturbances, then the theory of disturbance rejection control gives conditions and rules

for the design of a suitable control to decouple the interesting variable (2.8) from the disturbances, i.e. in the
closed loop system the disturbances do not effect the interesting variable any more. Particularly in the case if some
information on the type of the disturbance signals is available, e.g. step, ramp or harmonic function behaviour, a set
of linear time-invariant differential equations is used to
characterize the disturbances. By Müller and Lückel (5,6) a
detailed discussion of the disturbance rejection control was
presented showing how the actual disturbances will be estimated by a disturbance observer and how the estimates will
be fed back to compensate the influence of the disturbances
on the interesting variables.

   Therefore, the frictions (2.10, 2.11) operating in motor and gear of the drive are interpreted as "external
disturbances" which are almost piecewise constant:

$$r_1 \approx w_1 \ , \ \dot{w}_1 = 0 \ \text{ piecewise,} \tag{2.14}$$

$$r_2 \approx w_2 \ , \ \dot{w}_2 = 0 \ \text{ piecewise.} \tag{2.15}$$

Introducing an extended state vector

$$x_e = \begin{bmatrix} x \\ w \end{bmatrix} \ , \tag{2.16}$$

for the design of a disturbance rejection control the ficti-
tious system

$$\begin{bmatrix} \dot{x} \\ \dot{w} \end{bmatrix} = \begin{bmatrix} A & G \\ 0 & 0 \end{bmatrix} \begin{bmatrix} x \\ w \end{bmatrix} + \begin{bmatrix} b \\ 0 \end{bmatrix} u, \tag{2.17}$$

$$y = \begin{bmatrix} C & 0 \end{bmatrix} \begin{bmatrix} x \\ w \end{bmatrix} \ , \tag{2.18}$$

$$z = \begin{bmatrix} f^T & 0 \end{bmatrix} \begin{bmatrix} x \\ w \end{bmatrix} \tag{2.19}$$

is considered. According to (5,6) in a first step it is
looked for a linar feedback

$$u = - k_x^T x - k_w^T w \tag{2.20}$$

which yields a stable closed loop control system and de-
couples the intersting variable z from the signals w. It
can be shown that for (2.17, 2.19) such a control exists
where the design has to satisfy following conditions:

k_x^T can be designed by standard methods of
state feedback such as (2.13),   (2.21)

$$k_w^T = \frac{f^T (A - bk_x^T)^{-1} G}{f^T (A - bk_x^T)^{-1} b} \cdot$$ (2.22)

The determination of (2.19) results in

$$k_{w1} = - \frac{1}{K_M K_I} ,$$ (2.23)

$$k_{w2} = - \frac{1}{K_M K_I} - \frac{1}{c_{23}} k_{x1}$$ (2.24)

where the parameter were defined by (2.3 to 2.5) and k_{x1} is the first element of the gain vector (2.13).

In a second step the control (2.20) hast to be realized by feeding back the measurement (2.6). Even if the first part of the control (2.20) is a static output feedback (2.12), for the second part of (2.20) an observer has to be included. But this "disturbance observer" is used only for the reconstruction $\hat{w}$ of the actual frictions. Making available the measurement y_5 (2.7) of the torque at the driven end of the gear to the observer, it is possible to decouple the joint from the connected arm. Thus an functional observer may be only designed with respect to the subsystem consisting of motor, gear and friction models (2.14, 2.15). It does not need any data of the elastic arm (and additional joints and arms). The main advantage of this design is the reconstruction $\hat{w}$ of the frictions in each joint of a multi-axes robot depart from the other ones.

Unfortunately, the analysis of observability of the system (2.17, 2.18) shows that the components w_1 and w_2 are not reconstructable separately but only the sum $w_1 + w_2$ can be estimated. Therefore, the disturbance rejection control

$$u = - k_x^T x - k_w^T \hat{w}$$ (2.25)

cannot be realized exactly. But taking into account that

$$k_{w2} \approx k_{w1} = - \frac{1}{K_M K_I}$$ (2.26)

because the torsional spring c_{23} between motor and gear is very stiff, the disturbance rejection is performed approximately by

$$k_w^T \hat{w} \approx - \frac{1}{K_M K_I} (\hat{w}_1 + \hat{w}_2).$$ (2.27)

By this the feedback control (2.25) is given by the static output feedback (2.12) and by the rejection feedback (2.27) of the estimated sum of frictions in motor and gear.

The closed loop system shows in the case of constant sticking friction r_{10} and r_{20} steady-state accuracy with respect to the end position (2.8) of the arm by the exact rejection control $k_w^T \hat{w}$. However, for the approximate rejection control (2.27) a steady-state inaccuracy remains:

$$z_\infty = - \frac{r_{20}}{ic_{23}} .$$ (2.28)

But in any case a limit cycle does not appear.

Simulation results of the proposed position control of a single robot axis including the compensation of the Coulomb friction were shown in (1). Summarizing the results, the functional observer is able to follow the real friction torque (and backlash behaviour as well) very fast. There are very small differences between actual and estimated nonlinearities according to the differences between real system and observer model. Nevertheless, the closed loop control system shows very good behaviour because of the disturbance rejection control. Untimely stops of motion are prevented and a very high stationary accuracy and a good damping of the elastic vibrations of the system is obtained. The remaining very small position error is caused by the mentioned differences and by the approximation (2.27). The main advantage of the disturbance rejection control in comparison with integral control turns out to be its faster dynamic behaviour. The reconstruction of the friction is not limited by the dynamic of the controller and therefore may be much faster. Therefore, the new proposed control algorithm including the compensation of friction is superior to common control loops.

3.   MULTI-AXES CONTROL

In principal the control system for a multi-axes robot will consist of two control levels (7). On a first level each axis is controlled by the multi-layer feedback control developed above. On a second level the nonlinear couplings among the various axes are compensated by a coordination algorithm according to the theory of hierarchical decentralized control systems. The main advantage of this control concept consists of its realization by application of several parallel processors. The time critical operations can be realized by fast signal processors while the coordination of the coupled arm motions are controlled by a coordination processor. This concept leads to a fast and highly accurate control of the elastic industrial robot.

Firstly, the authors suggested in the contribution (2) to apply the technique of adaptive model following control. Based on a suitable reference model of the dynamics of each robot axis the adaptation loop has to correct the strong influence of variable moments of inertia depending on the actual position of the robot arms. But during the design of a adaptive model following control a simplified control algorithm was developed by applying the technique of robust control design in multi-model problems, cf. for example Ackermann (8). Designing simultaneously the gains of the static output feedback (2.12) with respect to four different typical end positions of the robot a fixed tuning of the gains

can be found which guarantees stable behaviour of the complete robot in the whole region of operation, cf. Ackermann
(9). Because the disturbance rejection control of each axis
is not influenced by the nonlinear coupling among the various axes, the complete control system of the multi-axes robot consists of a robust decentralized control of each robot
arm including the decoupled compensation of Coulomb friction
and backlash in each joint. The independent control of each
axis is also assisted by new results of Kiriazov and Marinov
(10). The implementation of a coordination algorithm on a
second level of a hierarchical control system is avoided because the feedback control of the first level is robust with
respect to the influences of the nonlinear couplings.

     Using a experimental laboratory robot with elastic ele-
ments which is described in detail by Henrichfreise (11) the
theoretical and simulation results could be confirmed experi-
mentally by Ackermann (9). There the efficiency of the new
robust decentralized position control of industrial robots
is convincingly illustrated. The tracking of quite complicated robot trajectories is performed with very small overshooting and very small absolute errors. The static and dynamic performance of the robot is improved even if friction
and torsional elasticity in the joints and elasticity in the
links arise. The implementation of the new robust decentralized position control could be realized by presently available signal processors such as a TMS320 processor.

4.   CONCLUSIONS

     The presented concept of a new robust decentralized
control design leads to a fast and highly accurate position
control of elastic industrial robots compensating the effects
of friction and backlash. Untimely stops of motions or limit
cycles are avoided. The efficiency of the proposed control
system was demonstrated by simulations as well as by experimental results of a three-joint elastic robot with
three rotational degrees of freedom. The results show a considerable improvement of   accuracy and of the dynamical behaviour of an industrial robot.

ACKNOWLEDGEMENT

     The authors gratefully appreciate the financial support
of this research work by the Deutsche Forschungsgemeinschaft
(DFG) under grant no. Mu 448/6.

REFERENCES

1.   Ackermann, J., and Müller, P.C., 1986, 'Dynamical Be-
      haviour of Nonlinear Multibody Systems Due to Coulomb
      Friction and Backlash', Austrian Center for Producti-
      vity and Efficiency, Proc. International Symposium
      'Theory of Robots', Vienna, Austria.

2.   Müller, P.C., and Ackermann, J., 1986, 'Nichtlineare
      Regelung von elastischen Robotern', VDI-Berichte Nr.

598 'Steuerung und Regelung von Robotern', Düsseldorf, FRG.

3. Den Hartog, J.P. 1931, Trans. ASME, APM-53-9, 107-115.

4. Marui, E., and Kato, S., 1984, Trans. ASME, J. Dyn. Syst. Meas. Control, 106, 280-285.

5. Müller, P.C., and Lückel, J., 1977, Regelungstechnik, 25, 54-59.

6. Müller, P.C., and Lückel, J., 1977, J. Problems of Control and Inform. Theory, 6, 211-227.

7. Müller, P.C., and Ackermann, J., 1986, 'The Effect of Friction to the Position Control of Industrial Robots', Akademie der Wissenschaften der DDR, Proc. Interdynamics 85, S-Reihe Nr. 7, Karl-Marx-Stadt, GDR.

8. Ackermann, J., 1986, 'Robuste Regelung: Beispiele - Parameterraum-Verfahren', VDI/VDE-GMA-Bericht 11 'Robuste Regelung', Düsseldorf, FRG.

9. Ackermann, J., 1988, 'Positionsregelung reibungsbehafteter, elastischer Industrieroboter', Dissertation, Bergische Universität - GH Wuppertal, FRG.

10. Kiriazov, P., and Marinov, P., 1987, 'On the Independent Dynamics Controllability of Manipulator Systems', Euromech Coll. 229, 'Nonlinear Applied Dynamics', University of Stuttgart, FRG.

11. Henrichfreise, H., 1988, 'Aktive Schwingungsdämpfung an einem elastischen Knickarmroboter', Dissertation, Universität - GH Paderborn, FRG.

Chapter 20

Detecting and avoiding collisions between two robot arms in a common workspace

R. A. Basta, R. Mehrotra and M. R. Varanasi

1 INTRODUCTION

The use of two or more robots in a common workspace is essential to enhance the utilization of robots, increase productivity, and improve the versatility of potential applications. If two or more robots are used in a common workspace, they may become obstacles to each other and, therefore, motion planning must include detection and avoidance of collisions between them. The problems pertaining to multiple robot systems have recently attracted the attention of several researchers DuPourque et al (1), Freund and Hoyer (2), Canny (3), Tournassoud (4), Fortune et al (5), Basta (6), Roach and Boaz (7), and Lee and Lee (8).

The focus of this research is on collision-free motion of two robot arms in a common workspace. A collision-free motion is obtained by detecting collisions along the straight line trajectories of the robots and then replanning the paths and/or trajectories of one or both of the robots to avoid the collision. In this paper, a novel approach to collision detection is presented and the techniques to avoid collisions are briefly discussed. Collisions are restricted to be between the wrists of the two robots (which correspond to the upper three links of PUMA manipulators). A sphere model for the wrist (including the tool and any grasped object) is used because it is rotationally invariant and computationally efficient. Collisions are assumed never to occur between the beginning points or end points on the straight line paths Basta et al (9).

2 COLLISION DETECTION

The detection of a collision is accomplished by calculating the distance between the origins of the two spheres representing the wrists of the robots. A collision is said to occur between the two wrists at any given time instant if the distance between the center of the two spheres is less than or equal to r1+r2 where r1 and r2 are the radii of the two spheres. One possible method of detecting collisions is to check at carefully chosen discrete time instants if the two spheres collide. Lee and Lee (8) adopted this approach in which motion planning using such a collision detection technique had to be done off-line. The collision detection algorithm presented here involves two steps: 1) obtaining the potential collision path segments along the straight line trajectories without considering the motion characteristics, and 2) mapping the potential collision segment information into the time domain to obtain the space-time collisions.

2.1 Detecting Potential Collision Segments

Potential collision segments are the locations on the straight line paths of the robots where the possibility of collisions between the two robots exists without considering the motion characteristics of the two robots. Specifically, for each of the two straight line paths a segment is found where, for each point of that segment, there exists at least one point on the other segment which is less than or equal to r1+r2 distance apart. Since the ultimate objective is to avoid collisions, only the locations on each path where potential collisions begin and end are required to be determined.

Let the parametric equations of the straight lines representing the paths of the two robots be

$$P_1 = P_{1i} + \lambda(P_{1f} - P_{1i}) \qquad P_2 = P_{2i} + \gamma(P_{2f} - P_{2f}) \tag{1}$$

where $0 \le \lambda \le 1$ and $0 \le \gamma \le 1$. For a potential collision to occur,

$$\| P_1 - P_2 \| \le r1 + r2. \tag{2}$$

Another way of computing the potential collisions is to obtain the intersections of a straight line representing one of the two paths with the locus of the surface of a sphere of radius r1+r2 whose center moves along the straight line representing the other path. This is equivalent to expanding the radius of the sphere of one robot by the radius of the other sphere while shrinking the other sphere to a point. It is obvious in the case of straight line paths that there will be a continuous segment where the potential collisions exist.

In the standard cartesian coordinate system, a sphere of radius r=r1+r2 and center (x_c, y_c, z_c) is given by the equation

$$(x - x_c)^2 + (y - y_c)^2 + (z - z_c)^2 = r^2. \tag{3}$$

Letting the center of the sphere move along the path of robot 1, the straight line path of robot 1 can be parametrically defined as

$$x_c = x_{1i} + a_1\lambda, \qquad y_c = y_{1i} + b_1\lambda, \qquad z_c = z_{1i} + c_1\lambda \qquad 0 \le \lambda \le 1 \tag{4}$$
where
$$a_1 = x_{1f} - x_{1i}, \quad b_1 = y_{1f} - y_{1i}, \quad \text{and} \quad c_1 = z_{1f} - z_{1i}. \tag{5}$$

The straight line path of robot 2 can be expressed as

$$x = x_{2i} + a_2\gamma, \qquad y = y_{2i} + b_2\gamma, \qquad z = z_{2i} + c_2\gamma \qquad 0 \le \gamma \le 1 \tag{6}$$
where
$$a_2 = x_{2f} - x_{2i}, \quad b_2 = y_{2f} - y_{2i}, \quad \text{and} \quad c_2 = z_{2f} - z_{2i}. \tag{7}$$

Thus, seven simultaneous equations with eight unknowns are formed. Equations (3), (4), and (6) are combined to obtain

$$k_1\gamma^2 - 2k_2\gamma\lambda + k_3\lambda^2 + 2k_4\gamma - 2k_5\lambda + k_6 = 0 \tag{8}$$
where
$$k_1 = a_2^2 + b_2^2 + c_2^2, \qquad k_2 = a_1a_2 + b_1b_2 + c_1c_2,$$
$$k_3 = a_1^2 + b_1^2 + c_1^2, \qquad k_4 = k_xa_2 + k_yb_2 + k_zc_2,$$
$$k_5 = k_xa_1 + k_yb_1 + k_zc_1, \qquad k_6 = k_x^2 + k_y^2 + k_z^2 - r^2 \tag{9}$$
and
$$k_x = x_{2i} - x_{1i}, \quad k_y = y_{2i} - y_{1i}, \quad k_z = z_{2i} - z_{1i}, \tag{10}$$

which provides the positions along the straight line paths of robot 1 and robot 2 that are distance r apart.

For a given value of λ in equation (8), γ can possibly have zero, one, or two values. This is equivalent to obtaining no intersections, one intersection, or two intersections with the other path for a given location of the sphere on its path. In case of two intersections, the distance on the straight line path between the two intersection points is called the potential collision length. Thus, equation (8) represents the location of the intersection points with respect to λ (or vice versa with respect to γ) producing a parametric space potential collision region. Although equation (8) is a second-degree equation in two unknowns representing conic sections, the only possible potential collision regions that can be generated are points, lines and ellipses (9). The ellipse is the most common case and is shown in Fig. 1 on a Parametric-Space-Potential-Collision-Region Diagram (PSPCRD). From now on, the ellipse potential collision region is utilized in examples. Interested readers should refer to Basta et al (9) for details on the other cases. As stated earlier, only the segment on each path where potential collisions exist is required. The horizontal and vertical tangents of the ellipse provide the information for determining the four extreme values (λ_{icp}, γ_{icp}, λ_{fcp}, and γ_{fcp}) that represent the locations on each path where potential collisions begin and end. Notice that an ellipse can be generated with tangents outside the valid range of $0 \le \lambda \le 1$ and/or $0 \le \gamma \le 1$. These invalid tangents are produced by extreme collisions that do not occur on the paths of one or both of the robots. For such cases, finding the missing extreme values in the valid range is discussed in Basta et al (9).

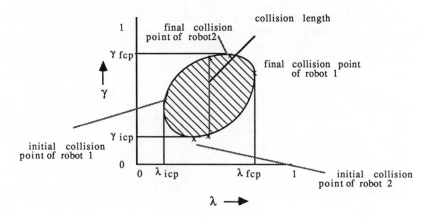

Fig. 1   Parametric-Space-Potential-Collision-Region
          Diagram (PSPCRD)

2.2 Detecting Space Time Collisions

Once the locations on each path where potential collisions begin and end are known (λ_{icp}, λ_{fcp}, γ_{icp}, γ_{fcp}), trajectory information can be used to determine whether a space-time collision (an actual collision in space and time) exists. If a space-time collision is likely to occur, it has to happen within the potential collision segment on each robot's path. Using trajectory information, the time range when the potential collisions along each path occur can be determined. Any overlap in the two time ranges suggests, but does not guarantee, the existence of a space-time collision. Therefore, an overlap in the time ranges when potential collisions occur is necessary, but not a sufficient condition, for determining if a space-time collision exists.

Thus, the method of detecting space-time collisions involves two steps: 1) determination of an overlap in the time ranges when potential collisions along each path occur to assure the possibilty that a space-time collision can happen, and then 2) confirmation of the existence of a space-time collision.

2.2.1 Determining Common Time Ranges. Using the trajectory information of a robot, the distance traveled along its straight line path with respect to time can be calculated. Letting distance traveled be defined by a parametric position, the following equations can be defined:

$$\lambda = \mathbf{f}_\lambda(t), \qquad \gamma = \mathbf{f}_\gamma(t) \tag{11}$$

and

$$t_\lambda = \mathbf{f}_\lambda^{-1}(\lambda), \qquad t_\gamma = \mathbf{f}_\gamma^{-1}(\gamma). \tag{12}$$

Equations (11) give the parametric positions on each path where the robots are located in time, and equations (12) perform the inverse which provides the time at which the robots reach specific locations along their paths.

Using the time equations (12), the time range when potential collisions along each path occur can be determined. A Space-Time- Collision-Region Diagram (STCRD) is shown in Fig. 2 which combines both path and trajectory information for a single break point case for each of the two trajectories. In this situation, an overlap in the time ranges does occur and a space-time collision region is formed. This region represents the positions along each path where a possiblity of a space-time collision exists.

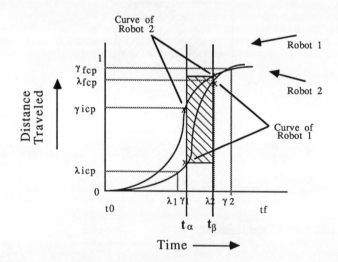

Fig. 2   Space-Time-Collision-Region Diagram (STCRD)

2.2.2 Establishing Existence of Space-Time Collisions. The main objective of collision detection is to establish whether or not a space-time collision occurs between the two robots. The collision region in the STCRD represents the positions along each path where a space-time collision can exist due to the overlap in the time ranges when potential collisions along the paths of the robots occur. The space-time collision region is delimited by the time of the initial potential collision of one of the two robots and by the time of the final potential collision of one of the two robots. The starting time and ending time of the region are denoted by t_α and t_β, respectively as shown in Fig. 2. This time range defines a curve in the STCRD for each robot. The curves represent the motion characteristics of the robots along their paths where space-time collisions are likely.

At a given time, the position of each robot can be determined by the position equations (11). Therefore, any time instant determines a (λ, γ) pair which can be transformed to another domain for analysis. This domain is shown in Fig. 3 as a Potential-Collision-Region-Motion Diagram (PCRMD). Thus, the two curves defined by t_α and t_β in the STCRD will result in one curve in the PCRMD. Therefore, the motion characteristics of each robot corresponding to possible space-time collisions can be analyzed with respect to the potential collision region. Let the starting point and the ending point of the curve in the PCRMD be defined as $C(\lambda(t_\alpha), \gamma(t_\alpha))$ and $C(\lambda(t_\beta), \gamma(t_\beta))$, respectively. The objective is to find whether or not this curve intersects the potential collision region. If the potential collision region is intersected by the motion curve, a space-time collision occurs since both robots are within colliding distance at some point in time defined by the (λ, γ) pair as shown in Fig. 3 as the true range of the space-time collision. An intersection with the potential collision region is determined by using the knowledge that since each of the functions in the STCRD is monotonically increasing, the (λ, γ) pairs that connect any two points on the curve in the PCRMD, such as the starting and ending points, must monotonically increase with respect to time. However, if the curve traverses from segment A, region 1 to segment D, region 1 or from segment B, region 3 to segment C, region 3, a space-time collision may or may not occur. Determination of the existence of a space-time collision for these two cases is accomplished by a fast iterative algorithm. The algorithm assumes that if the curve comes within some threshold distance from the potential collision region, a space-time collision occurs. In other words, the potential collision region can be thought of as being expanded slightly.

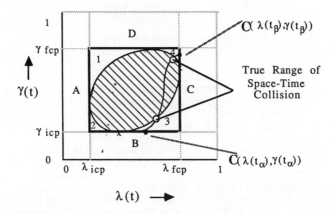

Fig. 3   Potential-Collision-Region-Motion Diagram (PCRMD)

The algorithm uses the fact that the (λ, γ) pairs defining the curve in the PCRMD monotonically increases with respect to time. The idea is to divide the curve into regions of bounding boxes such that the curve traversal within each bounding box is known. Fig. 4 illustrates the algorithm for the case where the curve does not intersect the potential collision region in region 3. Applying the approach to region 1 is straightforward. Basically, bounding boxes from the starting point to the ending point of the curve (left to right in λ and bottom to top in γ) are created by mapping between the STCRD and the PCRMD. At each iteration of generating a bounding box, only one new (λ,γ) point on the curve needs to be calculated since the previous (λ,γ) point is being used as the other point in the bounding box pair.

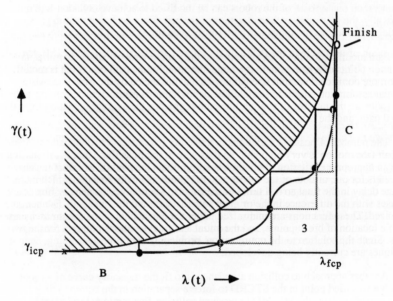

Fig. 4   Iterative Algorithm on a Collision-Free Path

However, if the curve intersects the potential collision region, an infinite amount of bounding boxes are generated each of which is decreasing in size as the curve approaches the potential collision region. Therefore, the iteration stops when the distance between potential collision region iteration positions is within some specified threshold. This is similar to expanding the potential collision region. The threshold defines a distance the curve has to be from the potential collision region in order to state that a space-time collision exists. The maximum possible distance occurs when the threshold value is produced by an equilateral triangle. Notice for these two cases that if a curve traverses near the potential collision region but never intersects the potential collision region, it is treated as a space-time collision although one never existed. If the iterative algorithnm is being used to determine the existence of space-time collisions, the acquired information can be used to reduce the space-time collision region for collision avoidance purposes. Precise details of the collision detection techniques are given in Basta et al (9).

3 COLLISION AVOIDANCE TECHNIQUES

To obtain collision-free motion, a collision in time and space is avoided by modifying the paths and/or trajectories of one or both of the robots. The method of achieving this collision-free motion depends on the avoidance requirements which modify various trajectory and path parameters. Due to space limitations, details are not presented here and interested readers should consult Basta et al (9).

A space-time collision region in the STCRD is depicted by common time ranges when potential collisions along the paths occur (or a reduced range when possible). The objective of collision avoidance is to eliminate the space-time collision region by producing non-overlapping time ranges. This is accomplished by modifying the paths and/or trajectories of one or both of the robots. Trajectory modification involves alteration of the motion characteristics of a robot along its path. Several parameters define the motion characteristics of a robot. These include number of break points, position of break points, chosen constant acceleration, and the starting time of motion. Any combination of the above parameters on one or both of the robots can be modified to achieve collision-free motion. Specifically, the new motion characteristics of the robots are such that the time ranges when potential collisions occur along each path (or reduced time ranges) do not coincide.

Path modification involves alteration of the path of a robot. Since the initial and destination points of a path remain the same, the straight line path of a robot is modified to two or more connected straight line paths which avoid the collision. This is consistent with the initial assumption requiring straight line paths. Various parameters describing a new travel route are the number of straight line path segments, the average deviation from the original path, and the new travel distance.

The number of possible variations in trajectory parameters or robot paths to achieve collision-free motion is very high. An obvious way to optimize these modifications is the use of an appropriate criterion to obtain a best solution. One method to compute these parameters for one or both of the robots is to minimize a penalty function which increases with the delay in the final arrival times of the robots. Alternatively, a penalty function which increases with the difference in the original path distance and the final path distance can be minimized. These functions are minimized under the constraints of maximum acceleration, possible location of break points, and the initial and final locations of travel for the two robots. Since the solution to the optimization problem is very time consuming, heuristic techniques are currently being developed.

Another approach to collision avoidance is to fit the trajectory curve of a robot through a specified point in the STCRD to force a separation in the common time ranges. A final approach utilizes the PCRMD to produce collision-free motion. The idea is to restrict the values of λ and γ to be within certain ranges such that a motion curve can never pass through the potential collision region. The motion curve can be modified within the bounding box of the potential collision region or it can be redefined from the start to bypass the potential collision region.

4 CONCLUSIONS

Collision-free motion of two robot arms in a common workspace is presented. The collision-free motion is obtained by detecting collisions along the straight line trajectories of each robot by using a sphere model for the wrists and then modifying the paths and/or trajectories of one or both robots to avoid the collision. The collision detection algorithm is described and approaches to collision avoidance are outlined. The efficiency of the collision detection algorithm allows for an on-line motion planner thus providing a significant contribution towards multiple arm coordination. Future research direction will encompass developing the heuristics for the collision avoidance techniques for two robots and generalize the techniques of detection and avoidance of collisions to handle more than two robots in a common workspace.

ACKNOWLEDGEMENTS

Support for this research by the NASA-Langley Research Center under Grants #NAG-1-632 and #NAG-1-772 and the AT&T Foundation is gratefully acknowledged.

REFERENCES

1.   DuPourque V., Guiot H., and Ishacian O., 1986, 'Towards Multi-Processor and Multi-Robot Controllers', Proc. IEEE Int. Conf. Robotics and Auto., 2, 864-870.

2.   Freund, E. and Hoyer, H., 1986, 'Pathfinding in Multi-Robot Systems: Solutions · and Applications', Proc. IEEE Int. Conf. Robotics and Auto., 1, 103-111.

3.   Canny, J., 1986, 'Collision Detection for Moving Polyhedra', IEEE Trans. PAMI, 8, 200-209.

4.   Tournassoud, P., 1986, 'A Strategy for Obstacle Avoidance and Its Application to Multi-Robot Systems', Proc. IEEE Int. Conf. Robotics and Auto., 2, 1224-1229.

5.   Fortune, S., Wilfong, G., and Yap, C., 1986, 'Coordinated Motion of Two Robot Arms', Proc. IEEE Int. Conf. Robotics and Auto., 2, 1216-1223.

6.   Basta, R., 1987, 'Multiple Arm Coordination Using Concurrent Processing', Proc. IEEE Southeastcon'87, 1, 328-331.

7.   Roach, J., and Boaz, M., 1985, 'Coordinating the Motions of Robot Arms in a Common Workspace', Proc. IEEE Int. Conf. Robotics and Auto., 494-499.

8.   Lee, B.H., and Lee, C.S.G., 1987, 'Collision-Free Motion Planning of Two Robots,' IEEE Trans. on SMC., 17, 21-32.

9.   Basta, R., Mehrotra, R., and Varanasi, M.R., 1987, 'Collision-Free Motion of Two Robot Arms in a Common Workspace', CSE-87-00002 Tech. Rep., Dept. of Comp. Sci. and Eng., University of South Florida, Tampa, FL.

Chapter 21

Work areas of manipulators—an improved algorithm

J. E. Sanguino, J. S. Mata, A. P. Abreu and J. J. Sentieiro

1. INTRODUCTION

Assume that the execution of a task requires the displacement of a particular tool along a certain surface with a very particular relative orientation. If this task is attributed to a robot manipulator, this must then be able to place the terminal effector on each point of the surface adopting the required orientation. It is then easy to understand that manipulators differing either on the number or type of joints, or on their ranges, will certainly exhibit different dexterities and consequently will be more adequate for different kinds of jobs.

In a recent paper by Yang and Chiueh (1) is described an algorithm capable of searching a work-area on any specified plane P for any specified hand orientation. Nevertheless, that algorithm has the disadvantage of being complicated for the user as regards the specification of the hand orientation relative to the desired plane P. Moreover, the complexity of the employed searching process should be avoided for the sake of simplicity and reliability when detecting voids and eventual separated work-regions of the work-area.

Having in mind the facts mentioned above, a simple, user friendly and more efficient algorithm is proposed taking advantage of a better and more intuitive definition of the plane P as well as of a set of simple rules. These provide the user with an easy and fast way of visualizing how to specify the manipulator's hand orientation.

Results of the application of this improved algorithm to a SCARA robot (three revolute and one prismatic joints) and to a PUMA 600 robot (six revolute joints) are shown.

2. WORKING PLANE AND MANIPULATOR'S HAND ORIENTATION

The desired plane P will be specified by the user in terms of the coordinate system {B} associated to the manipulator's base (frame of the link 0) through the corresponding plane equation, $Ax + By + Cz = D$. A frame {P} will be assigned to the plane P having its origin {P}$_{or}$ located at the point in P closest to {B}$_{or}$ and oriented according to the following rules : Zp axis orthogonal to the plane P, pointing outwards in relation to {B}; Xp axis horizontal (parallel to plane XoYo) and pointing to the right, for an observer looking from {B}$_{or}$; finally, Yp axis is obtained by applying the right hand rule from Zp to Xp. Notice that both Xp and Yp will be lying on the plane P since they are perpendicular to Zp. For a working plane parallel to XoYo, a frame {P} with the same orientation as {B} will be adopted. For a working plane parallel to XoYo the Xp axis must assume the same orientation as Xo.

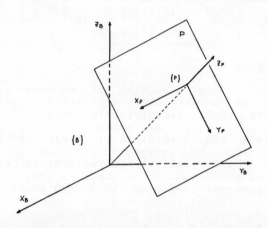

Fig.1   Position of the frame {P} relatively
to the frame {B}

Using this definition of the frame {P} , it is easy to set the corresponding homogeneous transformation $^{B}_{P}T$:

$$
^{B}_{P}T = \left[\begin{array}{ccc|c} & ^{B}_{P}R & & ^{B}P_{origP} \\ \hline 0 & 0 & 0 & 1 \end{array} \right] = \left[\begin{array}{ccc|c} Xp & Yp & Zp & ^{B}P_{origP} \\ \hline 0 & 0 & 0 & 1 \end{array} \right]
$$

where the matrix $^{B}_{P}R$ describes the orientation of {P} relatively to {B} and $^{B}P_{origP}$ is the vector that describes the position of {P}$_{or}$, i.e., the point in P closest to {B}, in terms of the frame {B}.

The evaluation of the Zp vector results from the specification of the plane P parameters, i.e., A, B, C, and D.

The Xp vector is defined by the cross product Zp x Zo, except if the working plane, P, is parallel to XoYo.

Finally Yp is obtained as the cross product of Zp and Xp.

The homogeneous transformation $_P^B T$ can now be put in the following forms :

A $\neq$ 0 or B $\neq$ 0

    D $\geqslant$ 0

$$_P^B T = \begin{bmatrix} B/S & (A.C)/(S.T) & A/T & (D.A)/T^2 \\ -A/S & (B.C)/(S.T) & B/T & (D.B)/T^2 \\ 0 & (-A^2-B^2)/(S.T) & C/T & (D.C)/T^2 \\ \hline 0 & 0 & 0 & 1 \end{bmatrix}$$

    D < 0

$$_P^B T = \begin{bmatrix} -B/S & (A.C)/(S.T) & -A/T & (D.A)/T^2 \\ A/S & (B.C)/(S.T) & -B/T & (D.B)/T^2 \\ 0 & (-A^2-B^2)/(S.T) & -C/T & (D.C)/T^2 \\ \hline 0 & 0 & 0 & 1 \end{bmatrix}$$

A = 0 and B = 0

    D $\geqslant$ 0

$$_P^B T = \begin{bmatrix} 1 & 0 & 1 & 0 \\ 0 & 1 & 0 & 0 \\ 0 & 0 & 1 & D/C \\ \hline 0 & 0 & 0 & 1 \end{bmatrix}$$

    D < 0

$$_P^B T = \begin{bmatrix} 1 & 0 & 0 & 0 \\ 0 & -1 & 0 & 0 \\ 0 & 0 & -1 & D/C \\ \hline 0 & 0 & 0 & 1 \end{bmatrix}$$

$$S = \sqrt{A^2 + B^2} \quad , \quad T = \sqrt{A^2 + B^2 + C^2}$$

Notice that, thanks to the assumed rules, ^B_PT is easily built. In fact the third and fourth columns, represent respectively the vector orthogonal to the plane P and the position vector of its closest point to the origin. The remaining columns are computed as cross products of known vectors as defined above.

This explicit way of defining the plane's frame orientation facilitates the user on his task of specifying the orientation of the manipulator's hand frame {H}, which must be done relatively to {P} by finding the rotation matrix ^P_HR:

$$^P_HR = \left[\begin{array}{c|c|c} Xp & Yp & Zp \end{array} \right]$$

Note that only six of its nine elements must be specified, resulting the third from their cross product.

3. POLARITIES OF A MANIPULATOR

A manipulator having N degrees of freedom might reach a point inside its work-space adopting up to $2**(N-1)$ different configurations, corresponding to the $2**(N-1)$ solutions of the inverse kinematic problem. This number is greatly reduced if a particular arm structure, limited joint ranges and particular hand orientation are imposed. As an example the PUMA 600 won't be able to adopt more than eight configurations, also called polarities, while the SCARA will be limited to only two .

4. IMPLEMENTATION OF THE ALGORITHM

After having been specified both the plane and the hand orientation, the search, regarding work-area definition, will be made on the given plane (here it will be taken advantage of the location of the coordinate frame {P}).

Since {P}$_{or}$ is at the nearest point from {B}$_{or}$, the coordinate frame {P} should be in the centre of the searching area. If the manipulator's workspace was a sphere with the centre in {B}$_{or}$, the intersection of the work-sphere with a given plane would be a circle with {P}$_{or}$ as centre. So, if r is the radius of the sphere that envelops the whole of the manipulator workspace, the searching area can be limited to a circle whose radius is r and the centre is in {P}$_{or}$, or else to a 2r side square (in this case, easier to implement). If the manipulator's structure is known, the value of r is always easy to compute.

In order to guarantee good graphic resolution the search on the square is made over a great number of points. For each of them, the inverse kinematic problem must be solved in order to investigate if there exists a possible solution. This means that the manipulator's hand frame {H} must be knowm in terms of the frame {B} for each point.

Two steps have to be taken in order to compute the transformation $^B_H T$, that defines {H} in terms of {B}.

First, the hand frame {H} must be known in terms of the plane frame {P}. Note that when the manipulator's hand is placed on the plane P then {H}$_{or}$ belongs to it. So the frame {H} can be related to the frame {P} considering both the hand orientation, specified by the user, and the coordinates $(p_x, p_y, 0)$ of the chosen point. Therefore the relationship of the hand about the plane, denoted by $^P_H T$, can be written as:

$$
^P_H T = \begin{bmatrix} & & & \vdots & p_x \\ & ^P_H R & & \vdots & p_y \\ & & & \vdots & 0 \\ - & - & - & - & -\!-\!- \\ 0 & 0 & 0 & \vdots & 1 \end{bmatrix}
$$

Second, the plane frame {P} must be known in terms of the frame {B}. The frame {P} is related to the frame {B}, by means of the plane definition, through the transformation $^B_P T$ above-mentioned.

According to these two steps the relationship of the manipulator's hand about the base can be represented as follows:

$$
^B_H T = {}^B_P T \cdot {}^P_H T
$$

Note that only the elements (1,4), (2,4) and (3,4) of $^B_H T$ depend on the point $(p_x, p_y, 0)$ chosen. So the remaining elements can be evaluated out of the searching process. This can increase greatly the quickness of the algorithm.

The flow chart of the algorithm is shown in Fig.2.

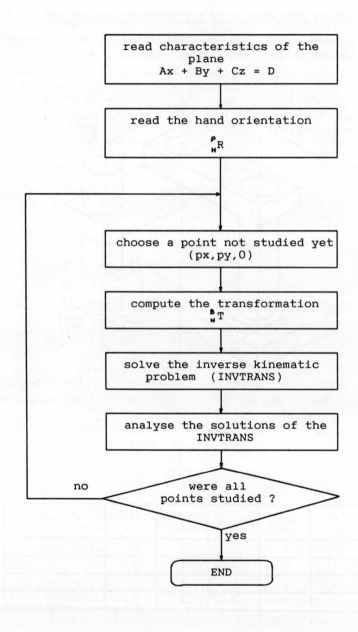

Fig.2   Flow-chart of the algorithm

5. RESULTS ACHIEVED WITH THE APPLICATION OF THE ALGORITHM

To illustrate applications of the algorithm, a SCARA and a PUMA (2),(3) robot were used (Fig.3 and 4).

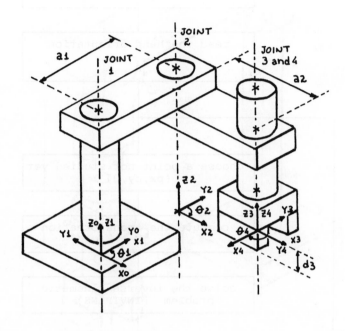

i	α_{i-1}	a_{i-1}	d_i	θ_i	range(o)
1	0	0	0	θ_1	$-135 \le \theta_1 \le 135$
2	0	140	0	θ_2	$-135 \le \theta_2 \le 135$
3	0	140	d3	0	$0 \le d3 \le 40$
4	0	0	0	θ_4	$-135 \le \theta_4 \le 135$

Fig.3   SCARA robot

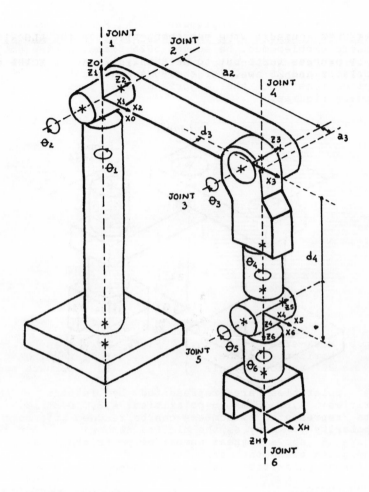

i	$\alpha(o)$ $i-1$	a(mm) $i-1$	d(mm) i	$\theta(o)$ i	range(o)
1	0	0	0	θ_1	$-160 \le \theta \le 160$
2	-90	433.5	0	θ_2	$-225 \le \theta \le 45$
3	0	19.1	125.8	θ_3	$-225 \le \theta \le 45$
4	-90	0	433.5	θ_4	$-170 \le \theta \le 170$
5	90	0	0	θ_5	$-135 \le \theta \le 135$
6	-90	0	0	θ_6	$-170 \le \theta \le 170$

Fig.4   PUMA robot

The algorithm was implemented in PASCAL. There, 5625 points were used in the searching area, meaning the inverse kinematic problem must be solved 5625 times. What might look a slow process turns out to need only 12 sec. for the SCARA manipulator and 25 sec. for the PUMA manipulator.

Applications of the algorithm are presented in the following figures.

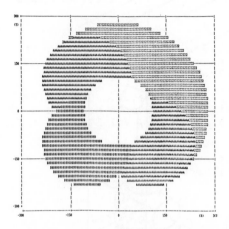
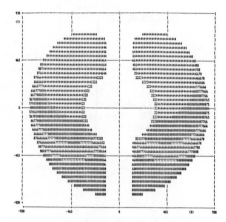

Fig.5   Work-area for SCARA, horizontal plane, 90° approach angle

Fig.6   Work-area for PUMA 600, vertical plane, 90° approach angle

The polarities are represented by letters (SCARA - polarities: A,B; PUMA - polarities: A,B,C,D,E,F,G,H). The points represented by numbers can be reached with more than one polarity (as many as the plotted number).

Figure 6 shows a great number of polarities meaning that in the given plane the manipulator has great dexterity.

Different polarities can be shown separately.

6, CONCLUSIONS

The present algorithm when compared with the one by Yang and Chiueh, reveals a better performance as regards computing time and simplicity of utilization.

The search process used is simple and presents the advantage of searching all the polarities simultaneously, which increases greatly the speed of the algorithm.

It must be underlined that this algorithm can be easily adapted to any manipulator, provided that the corresponding block of the inverse kinematics is substituted. Note also that the characteristics of the manipulator only influence the resolution of the inverse kinematics (INVTRANS Block).

This algorithm can also be useful as an aid in robot arm design. When designing new robot arm structures, it can be used to find the spaces and areas corresponding to different possible sets of arm parameters. Used in an iterative way it can help convergence to a design satisfying specifications expressed in terms of work-areas (depending on certain types of jobs - e.g. painting surfaces with certain orientations etc.)

It should also be pointed that it can have some importance as an instrument helping the potential user to decide which manipulator is better suited for a certain task.

7. REFERENCES

1. Yang, D.C., and Chiueh, T.S., 1986, 'Work-Area of Six-Joint Robot with Fixed Hand Orientation',Int. J. Rob. and Aut., Vol.1, no.1, 23-32.

2. Paul, R.P., Shimano, B., and Mayer, G.E., 1981, 'Kinematic Control Equations for Simple Manipulators' IEEE Trans. on Sys., Man and Cyb., Vol.SMC-11, no.6, 449-455.

3. Lee C.S.G., 1982, ' Robot Arm Kinematics Dynamics and Control ', Computer, Vol.15, no.12, 62-80.

Chapter 22

The development of the design criteria for the command input device for teleoperated robotic systems

R. M. Crowder

1. INTRODUCTION

A teleoperated robot will extend the capabilities of the human operator by allowing the manipulation of objects in environments that are remote and/or dangerous. The original teleoperator controller was based on the master/slave manipulator, where a master arm is coupled to a geometrically identical slave arm, placed in the remote environment. With the introduction of bilateral control strategies, as described by Hewitt and Siva (1), good positional and force control can be achieved. However, if the robot manipulator is not anthropomorphic in size or configuration the use of the master/slave concept becomes prone to difficulty.

The control of very large non-anthropomorphic manipulators, such as those described by Perratt (2), used in nuclear reactors, is of particular relevance to the Central Electricity Generating Board. Due to the large cost of operating these manipulators, any improvement in operator performance would be welcomed, preferably with improved safety. At present these manipulators are operated either on a joint-by-joint strategy or by resolved motion cartesian control. However these control techniques can be rendered ineffective if the relationship between the positional information displayed to the operator and the input controller's motion do not exhibit the 'axiom of naturalness'. Figure 1 shows an outline of a teleoperated system, as considered in this paper. The manipulator is controlled by an operator using a command input device. Operator feedback is via a number of sensors, the most important being vision. Due to the complexity of the manipulator's geometry, a computer is to resolve the input commands into joint positions.

The objective of the study reported in this paper was to review and then develop an understanding of the interaction between the operator, the manual input controller and the visual feedback in the teleoperated environment. Additionally the paper describes the design criterias that have to be observed in the development of command input devices in order to achieve satisfactory control of a teleoperated robot.

2. CONCEPTS OF TELEOPERATOR DESIGN

The objectives in the design of a teleoperated control system can be summarised as:

The minimisation of the operator's reaction time,
Ease of learning,
Correctness of any initial move,
Speed and precision of the control adjustment and,
Mininimisation of operator fatigue.

In a teleoperated system the operator will typically view the
manipulator via a closed circuit television system, and feed in his
commands via the main command input device. In the consideration of
teleoperated systems, it is the relationship between the image presented
to the operator and the direction of motion on the command input device
that is crucial to achieving satisfactory control.

The main factors in the evaluation of the operator's work load are
summarised in Figure 2, Vertut (3). Any evaluation must take into
account the aspects of the teleoperator system that are functionally
linked to the operator; the demands of the task, the operator's ability
and finally the system's performance.

Due to the complexity of the measurement of these parameters, it
is not easy to derive an overall measure of workload. The physical
loads can be determined, the mental workload being more difficult to
evaluate. However by the study of a number of human factor criteria the
operator's mental effort can be reduced.

2.1   Automatism, directional correspondence and transparency

The operator's mental workload will be reduced if he uses automatism, this being an action that is performed without the operator's
conscious knowledge, the key to achieving automatism is stimulus
response compatability. Stimulus response compatability requires that
the human response made must have a natural correspondence in both time
and space with the displayed actions or requirement. To succeed the
correspondence must be between the displayed action not the actual
manipulator's movement, and the motion of the command input device. If
this is not present, the operator will be required to mentally rotate
the movement he sees, before commanding the manipulator to move.
Vertut (3) showed that if the motion of the command input device is
rotated up to 30° relative to the displayed motion, the operator can
still achieve satisfactory control. If the phasing error is greater
than 45°, the operator experiences severe difficulties in controlling
the system.

The minimisation of the phasing errors in the operator's view
point will produce a transparent control system, where the operator
believes he is at the remote site and not decoupled by the teleoperator
system, allowing the operator to identify his own body and his immediate
environment with the manipulator and its environment.

A large number of simulations were conducted at the University of
Southampton, to determine the effect of phasing errors in teleoperated
systems. The task selected was a simple two dimensional tracking task,
simulated on a microcomputer. This allowed a number of command input
devices to be evaluated at different phasing errors. A typical result
for a rate controlled system is shown in Figure 3. The two main
characteristics are the initial error in movement, in the same direction
as the phasing error and, secondly, a tendency to circle around the
target in decreasing circles until contact is made.

To improve overall transparency the actual position of the display
compared to the command input device should match the position of the
remote camera compared to the manipulator arm. This correspondence

between the display and the body is not merely important, it is critical, especially in simple systems in an unstructured environment. This can be overcome by mounting the viewing system on the manipulator in line with the end effector. However this may not be satisfactory if the end effector becomes inverted, this will lead to a loss of the operator's vertical perception. The use of multiple camera systems will also lead to confusion, as the correct image may not readily be apparent. It is possible to switch automatically between viewing positions as a function of manipulator position, or link the kinematic joint solutions to the position of the camera relative to the command input device.

2.2   Minimisation of the degrees of freedom

The operator's workload can be reduced by restricting the number of degrees of freedom at the operator's command input device. The existing MEL manipulators use midrange control for translational movements in free space. The operator has only to control three translational degrees of freedom, since the manipulator has six degrees of freedom, three are redundant. Whilst the end effector is moved according to the operator's command, the redundant three joints are used to keep all the manipulator joints close as possible to their midrange. This is similar to the control methods proposed by Swain (4) for prosthetic limbs.

3.   DESIGN CRITERIA SPECIFIC TO THE COMMAND INPUT DEVICE

3.1   The input command

The commands issued by the operator can be either symbolic or analogic. Symbolic commands represent the required action, for example commands typed in at a keyboard. Analogic commands are those where the commanding motion is physically analogous to the required motion, for example a potentiometer. Achievement of stimulus-response compatibility is not automatically accomplished by analogic controls, nor is it impossible to accomplish with symbolic commands.

Analogic commands are most suitable for real time remote manipulation control and easily achieve stimulus response compatability. The operator's stimulus in this class of teleoperators is primarily visual with force feedback from the command input device. The operator should not have to resolve his control motion into components of movement specifically suitable to the command input device. He should directly make a command without having to know the way in which the command input device resolves the various degrees of freedom from that motion.

An analogic command input device operating as a rate controller should be centre-sprung to ensure that it returns to the 'no velocity' position when the operator removes his hand. Were this not the case, the manipulator would continue to move until it reached a joint limit or an obstacle. To improve operator control further the controller can have a breakout region where a suitable force has to be applied before the controller and hence the manipulator will move.

3.2   Cross-coupling of manipulator movements

The cross-coupling of movements may occur when the operator inputs a manipulator movement by operating the command input device, but inadvertently causes some other input to be made. Just as perception or

implementation errors can result in undesired manipulator movement, so
does command input device cross-coupling. Its cause, however, lies in
the design of the command input device or the dynamics of the teleoperator system rather than in operator errors.

Cross-coupling is likely in the conventional two axis joystick to
which has been added a third twist movement. To point the joystick in
different directions, the hand is rolled from side to side. As the hand
rolls, it tends to twist at the same time, the twist is inadvertent and
will cause unwanted motion of the remote manipulator.

The command input device must be ergonomically designed to ensure
that cross-coupling cannot occur. A command input device with N
degrees of freedom must take as its input N independent human movements,
all of which must satisfy the criteria of stimulus-response compatibility.

If the system exhibits a well-defined cross-coupling due to its
dynamics, then the manipulator's control computer can compensate for
the coupling, and produce error-free motion.

3.3   Ergonomic design of workstation

The workstation, including the command input device must be
designed for optimum operator performance; the controls within easy
reach and the display being seen without strain. Especially important
is that the workstation must be designed to avoid postures which
require considerable muscular effort to maintain.

The command input device itself should also be ergonomically
designed. For instance to encourage the appropriate grip, and to
ensure that the movements are neither too stiff nor too free. The
correct spring force in a joystick has been found to improve tracking
performance by 30%. The proper movement resistance contributes to the
control feel and therefore to operator performance. A qualitative feedback of the system state also occurs and unburdens the operator's visual
task. Well trained subjects on average choose a stiffness of 1.7N/cm
Kruger (5).

4.   THE IMPLEMENTATION OF THE COMMAND INPUT DEVICE   .

There is probably no ideal command input device that is suitable
for every manipulation system. The device chosen will depend on the
type or class of teleoperator system under consideration. For many
applications the master/slave controller will be the best solution to
command input. An excellent level of transparency is achieved, because
the command input device (the master) is a replica of the manipulator.
Indeed, tests show (3) that when feedback is good, the accuracy and
speed of this device, especially when augumented with force feedback,
exceeds any other method. Unfortunately, the master/slave configuration
is not suitable for every teleoperator system, including the type of
manipulator under consideration. The large remote arm would require a
scaled down master to make it of a sensible size. This in itself
presents no problems until a high degree of accuracy at the remote arm
is specified. When scaled down to the master arm, the accuracy
requirement may become smaller than the resolution of human movement.

It is possible to develop a number of command input devices that
are able to satisfy some or all of the human factor criteria specified
earlier. A number of different systems were reviewed during the work.
The control of large industrial manipulators was considered, hence

control techniques such as electromyographical systems as used on prosthesis, Nightingale (6), were not reviewed in depth.

The ergonomic controller is depicted in Figure 4. The diagram shows the correspondence between human arm, manipulator and command input device. It is a control input device which, it is claimed, is easy to use since it moves in the same way as a human forearm and hand. It requires a very tight coupling from operator to command input device. Unfortunately, a serious problem with this device is that not all the movements are compatible with the display information. If stimulus response compatability cannot be achieved, cross-coupling will occur, and degrade the system performance.

The joystick is probably the most common method of implementing a rate control command input device. The problem though is stimulus-response compatibility. Orientation control does not map well onto planar joystick motions, whilst the manipulator's translational motion does not map onto a joystick twist motion. Rate control enables the accuracy of 0.5mm to be obtained at the remote manipulator with a conveniently sized joystick.

Even with these problems the joystick has found wide acceptance as a control input device, since operators are adept at hand/eye co-ordination. In order to overcome the restrictions of orientation control a partitioning of the translational and orientation control is probably one of the most viable means of achieving complete manipulator control. A proposed orientation controller is a small model of the manipulator's end effector encapsulated within a sphere with its end point at the centre of the sphere. If this sphere is placed in some frame of reference, the orientation of the ball can be read with respect to the frame of reference and used to control the orientation of the end effector of the manipulator. The angle of rotation of the ball maps directly onto the angle of rotation of the end effector. Under position control, full movement of the sphere must be allowed to obtain all orientations of the end effector. This scheme satisfies the criteria outlined above, especially directional correspondence and naturalness. Problems arising from using a hand to control three motions on one device are not envisaged, because the hand has three independant rotations which also comply with the criteria of stimulus response compatibility.

Many other forms of command input devices have also been reported, some of which are suitable for a teleoperated robot, these include an instrumental glove with a location detector and multidegree of freedom T bar, Figure 5. In the latter, the bar is moved in any direction in the available space, both in translation and orientation, the displacements being determined by the connecting cables.

5.   CONCLUSIONS

The objective of the work was to understand the operation of the command input device for a teleoperated manipulator and determine the interaction between the operator and the system. It was concluded that:

The human is a vital component within the control loop of a teleoperated robot. Satisfactory control can only be achieved if the degree of 'naturalness' is large. Additionally the choice and design of a command input device, together with the presentation of positional information is critical to the satisfactory operation of the system.

6.   ACKNOWLEDGEMENTS

The author acknowledges the assistance of A. Lee, N. Taylor and J. Vernon in the preparation of this paper, together with Mr. D. Perratt and Mr. R. Noble of the C.E.G.B. Marchwood Engineering Laboratories.

REFERENCES

1.   Hewitt, J.E. and Siva, K.V., The Nuclear Engineer, 27, No.1 pp.23–26.

2.   Perratt, D., 1985, Nuclear Engineering International, August issue.

3.   Vertut, J., Papot, L., et al, 1973, Transactions of the American Nuclear Society, 17, p.510.

4.   Swain, I.D., 1982, Adaptive Control of an Arm Control, Ph.D. thesis, University of Southampton.

5.   Kruger, W., 1979, Proceedings of the Annual Conference on Manual Control, Wright Patterson Air Force Base, Report AFFDL–TR–66–138.

6.   Nightingale, J.M., Chapple, P., et al, 1987, Journal of Biomedical Engineering, 9, p.273–277.

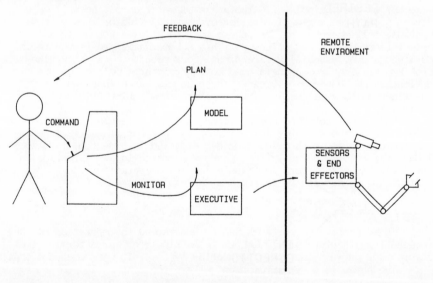

Fig 1. Outline of a teleoperated system.   The model and executive blocks are within the control computer.

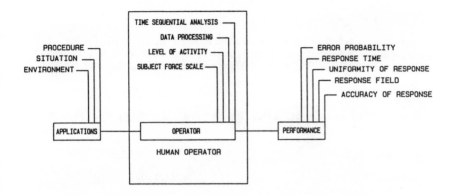

Fig 2. Operator's workload in a teleoperated system.

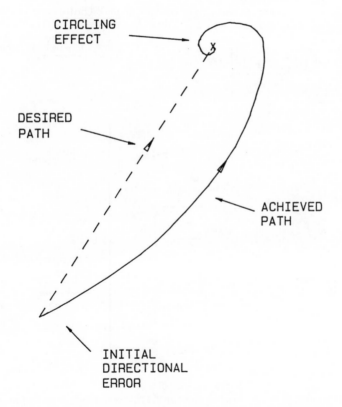

Fig 3. Typical tracking result, for a rate controlled application.

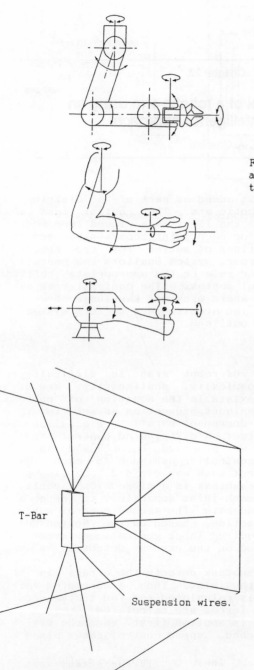

Fig 4.   Similarity between,
a manipulator, human arm and
the ergonomic controller.

Fig 5. T-Bar controller.
Movement of the T-Bar is
detected by measurement of the
tension in its nine suspension
wires

T-Bar

Suspension wires.

Chapter 23

Real-time control of a robotic arm using an artificial intelligence approach

A. Ghanem

1. ABSTRACT

An Expert System is added as part of a distributed control system for a robotic arm. The control task was divided into two subtasks: the first is a closed loop position and/or speed control for each joint separately, and the second solves the equations of motion for the required generalized forces. The expert system monitors the operation of the arm and requests or inferes the appropriate control gains for the first control subtask. The controller is also provided with a data base where previous knowledge about the control gains are stored. The architecture of this system is described and results are outlined.

2. INTRODUCTION

Real time control of robot arms is difficult to achieve because of the complexity, nonlinearity, and high degree of coupling that exists in the equation of motion. Distributed control techniques present an answer to this problem by achieving a degree of parallel processing to reduce the computation time. A distributed control system for a 3-joint robot arm was proposed before and applied successfully with a subcontroller assigned to each joint (Ghanem (1)). The control task was divided into two sub-tasks. The first control subtask is a closed loop position and/or speed control for each joint separately to force all joints to track the desired path. The second control subtask solves the equation of motion, based on the Newton-Euler formulation, for the required input generalized forces to apply to all joints based on the errors detected by the first control subtask.

The first control subtask detects the errors in the position and/or speed of the joint. Then, a corrected value of the acceleration ($\ddot{Q}_e$) is calculated and fed to the second control subtask where the forces are calculated. Maximum acceleration ($\ddot{Q}_m$) is applied first. When the desired maximum speed ($\dot{Q}_m$) is reached, speed control takes place by applying the acceleration:

$$\ddot{Q}_e = K_1 \ (\dot{Q}_m - \dot{Q}_a)/\dot{Q}_m$$, where $\dot{Q}_a$ is the actual speed.
Then, when the joint approaches the final position Q_f ,

deceleration starts by applying:
$$\ddot{Q}_e = -\ddot{Q}_m (\dot{Q}_a / \dot{Q}_m) + K_2 (Q_f - Q_a),$$ where Q_a is the actual position. A detailed description of this control strategy is found in (Ghanem (2)). This control strategy proved to be very efficient in guiding a robot arm through any desired path. Distributed processing helped in achieving the control task in real time. However, this control scheme did not eliminate the dependency on the operator who had to supply the controller with the control gains K_1 and K_2 for every path. Control gains are normally found by simulation, experimentation, or intuition. Artificial Intelligence technology presents an answer to this problem.

Artificial Intelligence techniques are intended to design computers that can imitate the human thought processes and robotics are prime candidates for it (Daniels (3)) Several researches have been reported in the literatures dealing with adding sensory capability and machine intelligence to the controls of a robot arm (Hall et al (4), Fulsang (5), Luo (6)). Others address the coupling between the robot and the physical world through path planning using knowledge-based systems (Gilmore et al (7)). In this paper, an expert system to infere the control gains is added to the distributed controller previously outlined. First, the architecture of the controller is introduced.

3.   THE ARCHITECTURE OF THE CONTROLLER

The distributed controller for a 3-joint arm consisted of three subcontrollers. Each subcontroller was implemented using the Intel's single board computer SBC 86/12A, with an 8087 NDP attached to the 8086. Subcontrollers communicated through a MultiBus. Although one of the subcontrollers started the control task, the system behaved as a multiprocessor with all processing elements having the same priority. Synchronization between the three processing elements was carried out by each subcontroller.(2)

To add an expert system to this controller, another processing element is included together with a memory module for the data base. Master-slave architecture is adopted for the system because it was found to be adequate for this application (Kanayama et al (8)) For a 3-joint arm, the architecture (as shown in Fig.1) has a master processing element, three subcontrollers, and a memory module. The master unit is implemented using another SBC 86/12A. All units communicate through a MultiBus. Each processing element has 32K of memory on board and the memory module has 64K.

The master, the supervisor, executes the expert system and monitors the operations of the subcontrollers. It is the only processing element that has access to the data base system in the memory module. It also provides subcontrollers with the initial and final positions, speeds, accelerations, and control gains. It directs subcontrollers to execute the first control subtask, where the system will act as SIMD machine; or to the second control subtask, where the system behaves as MIMD machine. In the first control subtask,

synchronization is taken care of by the supervisor.   In the
second control subtask, SBC's have equal priority.

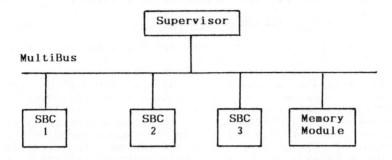

Fig. 1   The Architecture of the Controller

4.   THE EXPERT SYSTEM

The main functions of the expert system are to monitor
the operation of the arm and to request or infere the appropriate control gains for the first control subtask.   It also
maintains the data base where previous knowledge about the
control gains is stored. This knowledge is arranged in the
form of frames. Each frame contains information about the
initial and final points in the path, the moment and force
exerted by the gripper on the object (the load), the desired
maximum speed and acceleration, and the control gains for
the first control subtask.
The system operates in two modes of operation. In the
first mode, the system receives the control gains from the
operator or the path planning section. In this mode, the
expert system is transparent to the whole operation. It
passes these control gains to the subcontrollers and the
control task is carried out as mentioned before in (2). At
the same time, these values of the control gains are compared against the frames in the data base. If no match is
found, a new frame is generated and entered in the data base
for future use.
In the second mode of operation, control gains are not
provided to the expert system and inferring those control
gains is required. When the arm is about to move a certain
object, the expert system directs the gripper to exert the
appropriate moment and force to lift the object. These
values of moment and force are recorded as the load. Then,
the frames in the data base are searched for a close match
between this load and the entries of the frames. If a close
match is found, the corresponding control gains are picked
up. Then, the first and second control subtasks resume as
proposed before.
If a match between the load and the frames in the data
base is not found, the expert system attempts to infere the

control gains. Two inference strategies were tested. The first inferes the control gains using a simulation approach. The second corrects the control gains as needed.

4.1   THE FIRST INFERENCE APPROACH

In this approach, the expert system identifies two frames in the data base where the situation at hand falls in between. The values of the control gains in those frames are taken as the upper and lower limits of the desired control gains. Values between these limits are then picked and the performance of the control system is anticipated through simulation. If overshoots in the speed and/or position of any of the joints are detected in the simulation, other values of the control gains are picked. The process is repeated until an acceptable performance is achieved. A new frame is then created with the values of the control gains and entered in the data base for future use.

The simulation process requires a relatively long execution time. The control action cannot be delayed until the simulation process is completed and suitable control gains are found. Therefore, the two processes must execute concurrently if possible. However, the simulation task will must be completed before the control task is. A solution to this problem is to reduce the speeds of all joints which will increase the control time. This will allow the simulation to be completed.

To reduce the execution time of the simulation, the distributed processing approach is adopted. The simulation task is divided among the four processing elements in the system, the supervisor and the three subcontrollers. This has the effect of reducing the execution time of the simulation process to about one third.

The supervisor directs the subcontrollers to start the first and second control subtasks. At this point, the maximum speed and acceleration specified for the path are reduced by one half. Note that control gains are not needed in the first section of the first subtask. Then, between sampling times, the supervisor instructs all processing units to start the simulation task based on the values of the control gains picked up from the data base and also on the speeds and accelerations originally specified for the path. The timer on the supervisor board is used to interrupt the supervisor every 1/50s. When this occurs, the simulation task is interrupted and the control task is initiated again. Note that the upper limit of the sampling period for robot arm control is 1/60s (Luh et al (9)). This cycle repeats until the simulation task finds appropriate control gains. Then the supervisor will direct all subcontrollers to execute only the control task using the inferred control gains. Speeds and accelerations are boosted to the ones specified for the path.

The above scheme was programmed using assembly language. The actual 3-joint arm was not available and was simulated on a host computer interacting with the controller through the supervisor. It was found that this approach is

feasible   only if the control gains can be inferred in   less
than three iterations.   This occurs only when the data   base
contains   many frames with their entries not too far   apart.
In   many cases,   real-time control was not achieved and   the
system oscillated. Thus, the second approach was developed.

4.2   THE SECOND INFERENCE APPROACH

The   goal   of artificial intelligence is to   design   a
computer that imitates the human thinking process. Thus, the
expert system in this controller should imitate the way   the
operator   thinks   when   attempting to   predict   the   control
gains. If the operator is in control of the arm, he looks at
the load   and sees whether he had dealt with a similar   load
before   or not.   If the answer is yes,   he uses the   control
gains he used before.   If not,   then he will use the control
gains   known   to   him as a reference and try   some   out.   He
observes the performance of the arm. If overshoots occur, he
decreases   the control gains.   If the system is   overdamped,
then the control gains are increased. In short, the operator
uses his experience and some common sense.

This   is how the second inference approach attempts to
predict the control gains. The data base is first   searched
for a close match of the current load and the entries in the
frames.   If a match is found,   control gains are picked.   If
not,   then   two frames are identified as the upper and lower
limits   for the control gains.   The space between these   two
limits is divided into binary steps. Values in the middle of
these steps are then picked.

The   supervisor instructs the subcontrollers to   start
the control task using the picked control   gains.   Then,   it
will monitor the speed of each joint as well as the error in
position. Once an overshoot is detected, the control gain of
that   joint   is   modified by choosing another   gain   on   the
following   binary step towards the lower limit.   The subcontroller in charge of that joint is then istructed to use the
modified control gain.   This process takes place every sampling period.   When the joint reaches its final position, the
average   value of the control gain used for that   motion   is
entered in a new frame for future use.

The   strategy,   simple   in nature,   proved to be   very
adequate   to infere the control gains.   The overhead of   the
inference   engine   was   minimal and real   time   control   was
feasible.   Fig. 2 shows a comparison between the velocity of
the   second joint when the operator supplied the   supervisor
with wrong control gains: K_1 = 50, K^2 = 85,   and when   the
supervisor   inferred the control gains.   In this   path,   the
initial position was 0.00,   the final position was 1.00 rad,
and the maximum velocity was specified to be 1 rad/s. Fig. 3
shows the position of the second joint for both cases. It is
clear   that the inference rule is performing adequately   and
suggests that may be the inputs from the operator   should be
checked.

The time needed to search the data base depends on the
size   of   the   memory module and could add   to   the   overall
execution time of the controller.   However,   it was observed

that the supervisor was always able to find the limits of the control gains during the first section of the first control subtask where the gains are not needed.

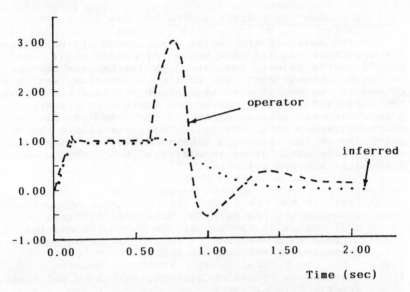

Fig. 2   The Velocity of the second joint

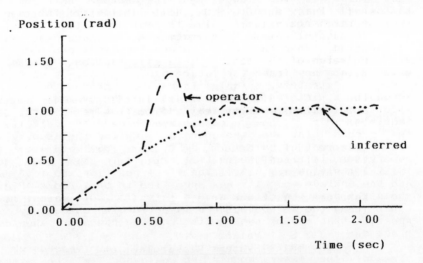

Fig. 3   The position of the second joint

5.   CONCLUSION

This paper presented an expert system added to a distributed controller for a robot arm to infere the control gains. A possible implementation for a 3-joint arm was outlined. Two approaches were presented. The first was based on evaluating the performance of the arm using simulation. This proved to be inadequate for real time control and slowed the speed of the motion. The second approach imitated the operator to infere the control gain. This was very successful in guiding the arm along the desired path.

The execution time of the controller without the supervisor was 3.460 msec when subcontrollers were driven by 5 MHz clock. With the supervisor, the execution time is 4.69 msec, which can be reduced by 50% if 10 MHz clocks are used. The difference in the execution time represents the execution time of the inference engine as well as the synchronization overhead. These results are adequate for real time control of a robot arm.

REFERENCES

1.   Ghanem, A., and Martens, H., 1986,'A Distributed Control System for a Robotic Arm', The 1st International Symposium on Robtics: Modeling, Teaching and Control, Albaquerque, New Mexico, U.S.A.

2.   Ghanem, A., 1986, 'The Control and Simulation of a Robotic Arm Using Microprocessor Based Distributed Control Approach', Ph.D. Dissertation, SUNY at Buffalo N.Y., U.S.A.

3.   Daniels, J.D., 1986, Signal (USA), 40, 10, 21-3,27

4.   Hall, E.L., and Oh, S.J., 1985,'Intelligent Robots for Factory Automation', Proc. SPIE Int. Soc. Opt. Eng. (USA), 548, 76-82. Arlington, VA, USA.

5.   Fulsang, E.J., 1985, Def. Electron. (USA), 17, 10, 77-82.

6.   Luo, R.C., 1985,'Artificial Intelligence in advanced Robotic Sensor Technology', Proc. of IECON 85, 1, 292-298.

7.   Gilmore ,J.F., Semeco, A.C., and Eamsherangkoon, P., 1985, Proc. SPIE Int. Soc. Opt. Eng. (USA), 548, 128-136. Arlington, VA, USA.

8.   Kanayama, Y., and Yuta, S., 1985, J. Robotic Syst. (USA) , 2, 3, 237-251.

9.   Luh, J.Y.S., Walker, M.W., and Paul, R.P., 1980, J. of Dynamics Systems Measurement and Control, 102,

Chapter 24

Human factors in the use of covariant bilateral manipulators

J. E. E. Sharpe

1. INTRODUCTION

Experimental evidence with instrumented simple hand tools has indicated that although the forward positional path requires a relatively limited frequency response of up to 20 or 30 hz, much of the sensory force information required to perform skilled operations is in the bandwidth up to 5-10kHz (1).

Recently a new class of bilateral control system that may be used in force feedback manipulators has been synthesised by applying Bond Graph methods (2). This is known as Covariant Bilateral Control. The use of covariant control provides a number of unique advantages over established methods of manipulator control, such as the widely adopted 'common error' system.

The use of the Covariant Bilateral Control system enables the Master and Slave systems of a manipulator to be designed independently, allowing the Slave positional servo to operate at the normal electrical or Hydraulic frequencies of between 15-30 Hz while the Master system which provides the feedback of sensory forces to the operator from the slave load or workpiece is designed to have a bandwidth greater than 300Hz and preferably 5-10 kHz, as shown in figure 1.

The results obtained from a 3 kW electro-hydraulic single degree of freedom system, have shown how important is the use of high frequency sensory force feedback in enabling an operator to control non-linear and resonant systems that cannot be controlled by conventional positional control. The use of feel enables the operator to adapt the level of damping within the overall system by altering his grip in response to the sensing of stick-slip and resonant vibration. The same feel enables the operator to sense gravity and inertial effects and these allow him to position the load more quickly and accurately than with conventional positional control.

Essentially, it is the ability to control a resonant system attached to the slave actuator outside the positional loop that demonstrates the importance of Covariant Bilateral Control. Tests have been undertaken in which the damping in the slave positional servo is reduced to nil. Under these conditions the operator with sensory force feedback automatically compensates and the overall system operates normally. Without force feedback the system is totally unstable under these conditions.

The introduction of time delays into a bilateral servomechanism poses a number of problems related to control stability and the fidelity and usefulness of the force reflection. Experimental work that has been undertaken on the Covariant Bilateral Electro-Hydraulic system controlling

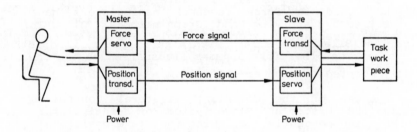

Fig.1   Electro-hydraulic   Covariant   Master-Slave Manipulator. Block Diagram

the position of a resonant load with adjustable delays in the forward and return paths, has demonstrated that a manipulator operating in earth orbit could conceivably be controlled with force reflection from an earth station, if a covariant system is used. The operator appears able to use the high frequency force feedback information in an adaptive manner. The master and slave systems rely on local control loops that are unaffected by the transmission delay and are therefore stable in their own right.

Covariant Bilateral Control of a number of different experimental 1 and 2 degree of freedom force reflecting systems have been shown to provide the operator with accurate sensory feel using a wide range of hardware and under different operational conditions. notable among these are the ability to use non-backdriveable servos and the control of resonant and variable inertia loads.

Experimental work especially with the system containing delay, demonstrates two different regimes of operator control. These are free systems where an inertia is to be positioned in space, in which the inertia dominates the performance, and stiff systems, in which a controlled force is to be applied into some structure. When positioning the inertia the introduction of the delay reduces the level of force feedback that can be used before the overall system goes unstable. In the case of the application of a force high levels of force feedback gain may be used. The Covariant Bilateral system with high frequency bandwidth of force sensing, therefore appears transparent to the operator which is the essential requirement of the system.

2.   THE COVARIANT BILATERAL MANIPULATOR SYSTEM

The Covariant Bilateral Electro-Hydraulic Manipulator used for these studies of human response was designed to the concepts outlined in the paper (1), presented to the Bond Graph session of the 1985 IMACS Symposium.

Covariant bilateral manipulator control was synthesised from an idealised Bond Graph structure representing the simplest form of manipulative tool. The ideal is to produce a system that is transparent to the 'operator'. In the covariant system the energy covariables at the operator and the workpiece are reconstituted by the Master and Slave systems respectively, as shown in figure 1.. The slave servo system uses a commercial 'vane' actuator controlled by a two stage hydraulic servo valve with reference to a positional potentiometer feedback, with damping provided by tachometer feedback.

The master control handle is mounted directly onto the output shaft of a DC servo motor and is fitted with force measuring strain gauges. The master control operates a

positional potentiometer to provide the input demand
position for the slave servo system.

The sensory force feedback to the operator is provided
by the output torque of the master servo motor which is
controlled with reference to the output torque of the slave
actuator as sensed by the differential pressure across the
actuator. The master servo has been carefully designed to
eliminate any difficulties due to the very fast response of
the system to the sensory forces.

The system has operated extremely well and has produced
a great deal of important quatitative evidence supporting
the view that the Covariant Bilateral Manipulator concept
has many very interesting properties and is capable of
providing very good 'Feel' even with difficult slave
actuators containing non-linearity, backlash and delay in
the positional and force reflecting path.

The force fidelity is a function of the quality of the
master system. It can faithfully reflect the effects of the
slave actuator seal friction and has a proportional response
to 270hz. Most important has been the ability, with force
feedback, for the operator to control the behaviour of a
resonant system attached to the output shaft of the Slave
actuator. Without this sensory feedback, the operator was
totally unable to control the system even when able to view
what was happening.

The output torque from the direct coupled servo motor
is limited. This appears to have a strange reversal effect
on the operator's feel, as though the operator responds more
to a change of force rather than actual steady forces. The
covariant system may use any combination of servo actuators
for either the Master or Slave. The Electro-Hydraulic
combination used for these studies was chosen deliberately
to introduce difficult non-linearities to establish the
limits of the system performance. However current work is in
hand with linear Master and Slave systems having a wide
bandwidthe in excess of 500 hz..

3.   HYDRAULIC SLAVE SERVO PERFORMANCE

The hydraulic servo uses a direct actuating vane type
of actuator. A single turn precision potentiometer is
directly mounted on the shaft to provide an output voltage
proportional to the actuator position. The hydraulic
pressures in the actuator are controlled by a two stage
'moog' control valve supplied by a 3KWe hydraulic power pack
operating at pressures up to 10 MPa. The gains of the
proportional amplifier and tacho-generator were adjusted to
give a slightly over damped response, with a natural
frequency of 20 hz. for small errors. The maximum flow

available from the power supply, however, restricts the
maximum velocity of the actuator to 200⁻/sec and gives the
non-linear reponse shown in the phase plane diagram of
figure 2.

The behaviour of the actuator seals have considerable
effect on the slave response by creating a pressure
dependant 'stick-slip' force on the actuator. Furthermore
the seal friction creates a pressure dependant hysteresis.
This 'stick-slip' tends to destabilise the system requiring
extra damping. It also excites any resonance in the load
attached to the actuator.

Tests were initially undertaken with a minimum inertia
directly coupled to the output shaft. However, in order to
provide a severe test for the overall system a flexible bar
was attached to the actuator output. This bar was fitted
with varying mass positioned at different radii. In moving
from one position to another the seal 'stick-slip' excited
the resonance of the mass on the fexible bar which the
control system was unable to control,as it was outside the
normal positional feedback loop of the controller.

The differential pressure across the actuator was
measured by a pair of pressure transducers operating into
strain gauge amplifiers and a difference amplifier. In
addition to providing the basic signals for the sensory
force feedback to the operator, these pressure measurements
enable the effects of the actuator seal friction to be
assessed.

The behaviour of this system is good but considerably
influenced by the non-linearities, especially the seal
friction. The slave system provided a very demanding system
to interface the electrical master to.

4.   THE ELECTRICAL MASTER SYSTEM

The master servo operates with its motor current being
controlled to be proportional to the sensory force feedback
signal derived from the differential pressure measurement
made on the slave actuator. The joystick is provided with
strain gauges to monitor the force exerted by the motor on
the operator.

The response of the motor with current control was good
to 270 hz. However, the maximum torque that could be
produced was limited to approximately 8 Nm. because of the
limit on the maximum stalled current output from the power
amplifier. Although this limited current output does not
matter in the overall performance of the system, it does
introduce some interesting effects which influence the
operator's perception, which will be discussed later.

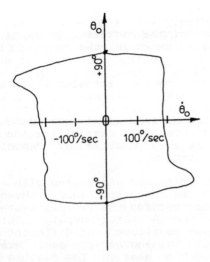

Fig. 2 Phase plane of slave actuator

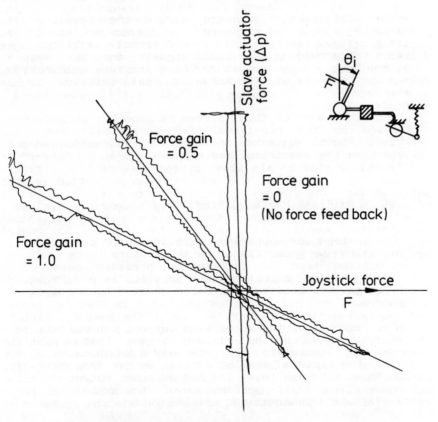

Fig. 3 Relationship of slave force and joystick
force showing force fidelity

The use of strain gauges to measure the bending moment
on the joystick created difficulties as it was possible for
the operator to introduce moments independant   of any
tangential force, thus the operator was unsure of what he
was really feeling.   Shear transducers that are unaffected
by the application of a superimposed moment are now being
used.

The relationship between the desired feedback force and
the force exerted on the joystick was linear up to the
limiting force due to the amplifier saturation.

5.   SENSORY FORCE FIDELITY

The sucessful operation of any sensory force feedback
system is dependant on the accurate reflection of the forces
acting on the slave manipulator.   In the system under test
the forces acting on the slave actuator including stick slip
and hysteresis have been faithfully transmitted to the
operator.   Figure 3. shows the relationship between the
force on the slave arm against the forces produced at the
joystick of the master as monitored by the strain gauges
fitted to the joystick.   Considerable care was taken to
apply the load slowly and deliberately and to avoid any seal
movement.   The response shows a clear proportional response
to the point where the power amplifier of the master servo
saturates.

It was preferable to measure the force reflected onto
the operator by measuring the slave servo motor current Im
and to relate this to the slave performance or load. Figure
4. shows the record of the reflected force plotted against
position for the slave swinging a mass of 4 Kg. at a radius
of 0.75 m. through an angle of +- 180 degrees against a
fixed stop. The record shows clearly the sinusoidal
gravitational force as the mass swings from the horizontal
through the vertical and up to the horizontal where it comes
against the fixed stop.

The fixed stop is reflected as an abrupt increase in
the reflected force limited only by the amplifier
saturation. The stick-slip of the seals is clearly evident
and the substantial hysteresis due to the seal friction is
shown to be a function of the differential pressure across
the seal.   Figure 5. shows a similar record taken with the
mass removed from the bar. The self weight of the bar
produces a reduced sinusoidal force, while the stick-slip
forces have a higher frequency. The comparison of these
records confirms the high degree of fidelity in the force
reflection and the ability of the system and the operator to

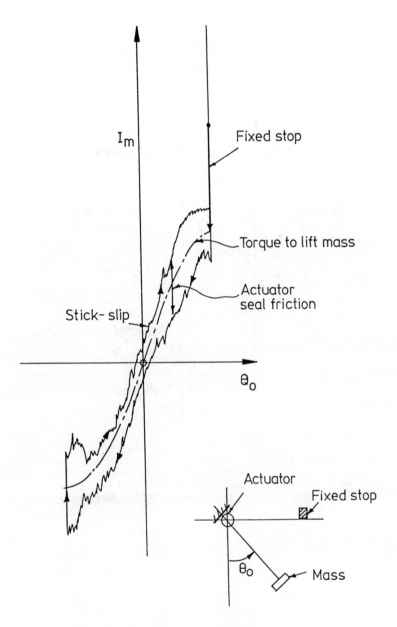

Fig. 4 Record of master servo motor current I_m
whilst controlling slave actuator with mass
through 180° arc and against fixed stop

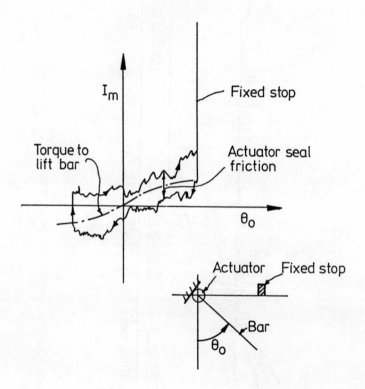

Fig. 5   Record of I_m whilst controlling slave
actuator and bar through 180° to fixed stop

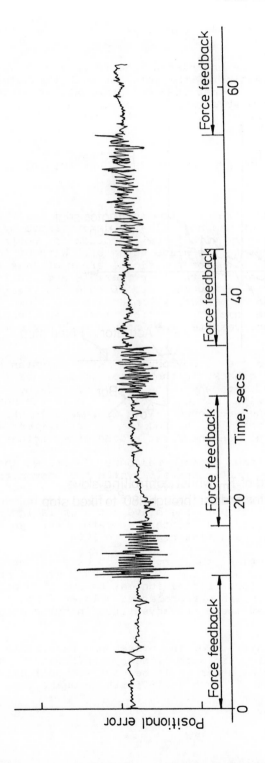

Fig. 6   The effect of force feedback on operator control of resonant
load attached to slave actuator

adapt to the changing mass parameter and the changed
dynamic situation.

Further tests were undertaken recording the joystick
force against the position of the actuator with the actuator
with no load moving against its own internal stop. These
also demonstrated the abrupt nature of the reflected force
at the stop and the repeatability of the system.

6.   OVERALL OPERATOR RESPONSE WITH COVARIANT BILATERAL SYSTEM

The overall response of the total system with and
without force feedback was most clearly demonstrated by the
ability of the operator with force feedback to control a
resonant mass attached to the slave actuator. Without force
feedback, despite an actuator positional error of less than
0.2%, the attached mass on the long flexible arm continued
to vibrate with little if any damping, as shown in figure 6.
Switching on the force feedback immediately allows the
operator to feel the small vibration level and rapidly
dampen the vibrations despite the frequency being some 10
hz..

Figure 7. shows the step response of the system with
and without force feedback. Without the force feedback the
resonant mass is excited when the actuator reaches the
desired position. With force feedback there is the normal
single overshoot damped response. This is remakable when it
is realised that this is a fourth order system with
positional control only on the first second order actuator.

In addition to the dynamic response described above,
tests were conducted on the overall system that demonstrate
the fidelity of the force feedback and the effect of
changing the force feedback gain. If the hysteresis due to
the internal seal friction in the slave actuator is ignored,
the sensory forces feedback to the operator through the
joystick have a linear and flat reponse to 120hz.

The act of changing the force feedback gain has no
effect on the dynamic behaviour of the overall system or on
its stability. The force reflection may be switched on and
off at will. Some operators preferred a lower force level
and it is clear that there are advantages in being able to
adjust the force level.

The overall behaviour of the system was remarkable for
its ability to provide so much sensory information with
bandwidth up to 270 hz which could be used subconciously by
the operator to control such a difficult resonant load and
non linear actuator.

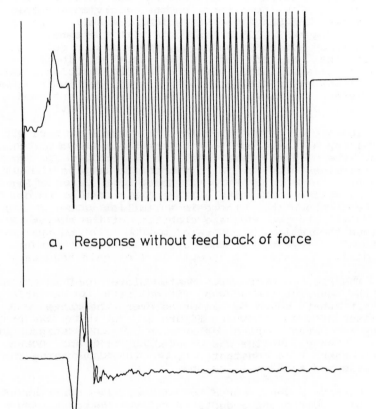

a, Response without feed back of force

b, Response with feed back of force

Fig. 7 Slave responses with and without
force feed back to operator

7.   OVERALL OPERATOR   RESPONSE OF   SYSTEM WITH TRANSMISSION
DELAY

The introduction of a pure transmission delay into the
forward and return paths between the Master and Slave
manipulators has no effect on the individual behaviour of
the Master and Slave servos which are self contained, as
shown in Figure 8. This contrasts with the behaviour of the
Common Error sytems in which any transmission delay would be
within the Master and Slave control loops. At the
frequencies associated with the forward positional path the
overall system is dominated by the mass that is to be moved.
The Master and Slave servos feel 'transparent' and may be
ignored in considering the overall system behaviour.

The system simplifies to one comprising a single loop
containing a lag, representing the Slave servo and inertial
load, the fixed gain of the operator K_{op} and the
Transmission Delay t. Also within the loop is the variable
gain of the Sensory Force feedback, as shown in Figure 9.
For any system containing a lag and a transmission delay, it
has been shown that there is a limiting value of loop gain
that will provide adequate stability of the system (2). The
maximum theoretical loop gain may be plotted against the
ratio of delay ,t, to the system lag time comstant,t. As
the delay increases the acceptable loop gain reduces.

The Electro-hydraulic Master Slave configuration used
for the earlier tests was fitted with a variable pure
delay,t, that could be adjusted over the range 2.0 - 180
milliseconds, as shown in figure 8. The delay was produced
using a charge coupled delay line of 3328 stages with 6
taps.   The delay line was clocked at 10kHz and 100kHz. The
first order time constant, t, of the Slave servo and load
was measured as 32 milliseconds.

Operators were asked to control the position of the
mass with different amounts of delay. Up to a delay of 6.0
millisecs the full force feedback loopgain could be used.
However for delays greater than 6.0 millisecs the acceptable
gain setting fell as the delay was increased, as shown in
figure 10.   The problem with reducing the loop gain is that
the force levels also reduce to such an extent that with a
delay of 120millisecs the forces tend to fall below an
acceptable threshold.   Results are plotted for 6 operators.

The effect of the delay is very clear to the operator
but the amount of the delay is very difficult to quantify.
This implies that the operators response is instinctive. A
comparison of the system operating with no delay and with a
delay of 20millisecs and a reduced gain is shown in figure
11.   This shows the reflected sensory force on the joystick
as the operator swings the mass through 180° against a fixed
stop.   The stick-slip vibrations are slightly worse with the
delay but the effect of seal friction is reduced.   The

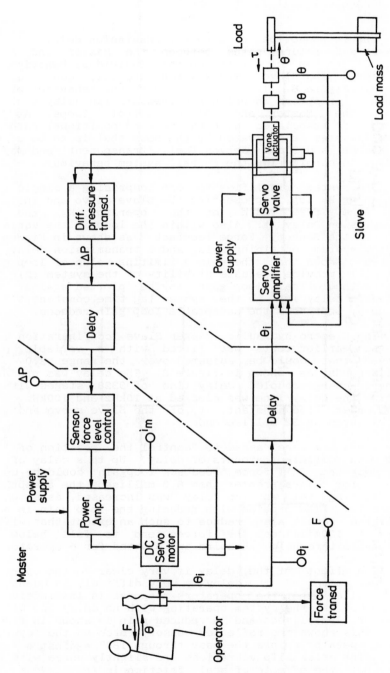

Fig. 8 Block Diagram of Electro-hydraulic Covariant
Master-Slave System with variable Transmission Delay

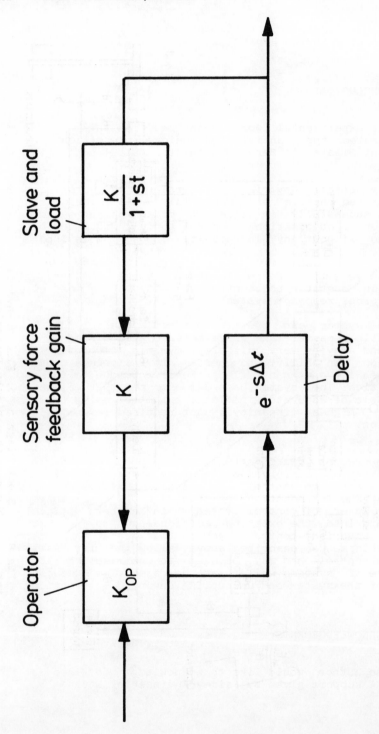

Fig.9   Simplified Block Diagram of System showing
1st order Lag and Delay

performance is quite acceptable especially in view of the
non-linearity in the system.

8.   CONCLUSIONS

The experimental work described in this paper has
demonstrated the anticipated unique properties of the
Covariant Bilateral Servo Manipulator albeit only in one
axis.

A practical Covariant Bilateral Servo system has been
demonstrated capable of industrial application that can
operate sucessfully in different modes and can be switched
from open loop positional control to full force feedback at
will and without any instability. The system is inherently
stable.

The very fast force feedback response (270 hz) allows
the human operator to control an otherwise uncontrollable
higher order resonant system.

The system can operate perfectly sucesssfully with
incompatible master and slave actuators of different energy
forms and having very different dynamic responses. It has
also shown an ability to operate in the presence of gross
non-linearity and can reflect the effects due to those nonlinearities. The system has demonstrated that although it
is capable of accurately reflecting the forces in the system
the very highest linearity is not required for satisfactory
operation.

In recent years attention has been given to the
development of sensory robots using a range of optical and
force sensors to control the overall performance of the
robot. The introduction of the sensors has introduced
another layer of control system which has been difficult to
integrate into the existing positional control system. The
use of Covariant control systems that are at the centre of
the developments described above allow the integration of
sensory control and may be used to optimise the dynamic
response of systems whilst at the same time reducing the
order of the system and uncoupling unwanted interactions.

9.   ACKNOWLEDGEMENTS

The author would like to acknowledge the financial and
material support given by Fairey Engineering Ltd.

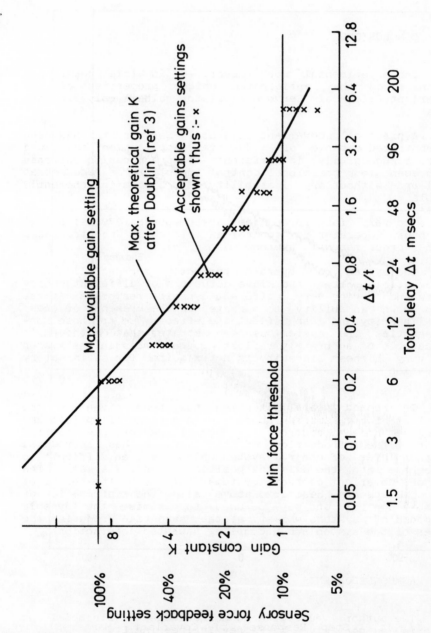

Fig.10   Relationship of Gain-Delay for theoretical
and practical system

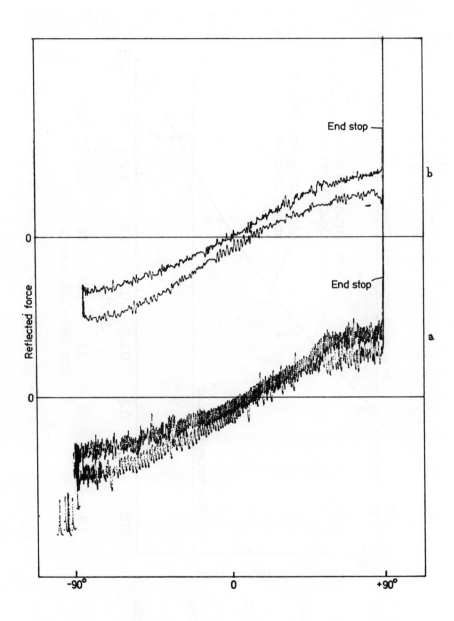

Fig.11   Step   Response   of   Covariant   Master-Slave
system, (a) with and (b) without delay

10. REFERENCES
1.   J E E Sharpe,   "Application of Bond Graphs to the
Synthesis and Analysis of Telechirs and Robots". 3rd CISM
IFToMM Conf   on Theory   and Practice of Robotics and
Manipulators, Udine, 1978.

2.   J E E Sharpe, K V Siva.   "Bond Graph Synthesis and
Analysis of Covariant Bilateral Servo Systems in Manipulator
Control". IMACS TRANS Vol 4   North Holland Netherlands
1986.

3.   E O Doebelin, "Proportional Control of a First Order
System with Dead Time". chap 9.4   'Control System Principles
and Design', Wiley 1986.

Index